BULLETIN OF THE OHIO BIOLOGICAL SURVEY
New Series

Volume 8   Number 2

ECTOPROCT BRYOZOANS
OF OHIO

By

TIMOTHY S. WOOD

Department of Biological Sciences
Wright State University

PUBLISHED BY
COLLEGE OF BIOLOGICAL SCIENCES
THE OHIO STATE UNIVERSITY

COLUMBUS, OHIO 43210   1989

OHIO BIOLOGICAL SURVEY

BULLETIN SERIES: ISSN 0078-3994
BULLETIN NEW SERIES VOLUME 8 NUMBER 2:
ISBN 0-86727-105-1
LIBRARY OF CONGRESS NUMBER: 87-072906

EDITOR

Veda M. Cafazzo

EDITORIAL COMMITTEE

CITATION

Wood, Timothy S. 1989. Ectoproct Bryozoans of Ohio. Ohio Biol. Surv. Bull. New Series Vol. 8 No. 2. x + 70 p.

COVER PHOTOGRAPH—Young colony of *Lophopodella carteri* from a laboratory reared population. Photograph by Timothy S. Wood.

COLLEGE OF BIOLOGICAL SCIENCES
THE OHIO STATE UNIVERSITY
COLUMBUS, OHIO 43210
1989

12-89—1.5M

ABSTRACT

A survey of 60 public access lakes, rivers, and streams throughout the state of Ohio has revealed at least 13 species of ectoproct bryozoans. The most common and widely distributed of these are *Fredericella indica, Plumatella casmiana, P. repens, P. emarginata, Hyalinella punctata,* and *Pectinatella magnifica.* Species reported from fewer than three sites in the state include *Fredericella australiensis, Lophopodella carteri, Cristatella mucedo, Paludicella articulata,* and *Pottsiella erecta. Plumatella reticulata,* occurring at eight locations, is only recently-described, and has not been reported outside Ohio. The western basin of Lake Erie alone has at least 10 of Ohio's bryozoan species. A checklist of freshwater bryozoan species in North America is included, as well as a key to freshwater bryozoans of the Great Lakes region. Collecting and preserving methods, and the rearing of bryozoan colonies in the laboratory are outlined. There is a brief discussion of the basic and reproductive biology of bryozoans. The taxonomy and biology of Ohio's species are provided, and for some, measurements are tabulated and analyzed.

Mary Dora Rogick

DEDICATION

This book is dedicated to the memory of Mary Dora Rogick (6 November 1906–25 October 1964). Dr. Rogick was the first serious student of Bryozoa in Ohio, and she made significant contributions throughout her life to the knowledge of this invertebrate group.

Born to Croatian immigrant parents in East Sandy, Pennsylvania, Mary Rogick was the only one of four children to survive beyond infancy. She attended public schools in Council Bluffs, Iowa, and went on to earn Bachelors and Masters degrees in zoology at the University of Nebraska.

In 1931, Mary Rogick enrolled in the doctoral program at The Ohio State University. She took summer courses at the university's Franz Theodore Stone Laboratory at Put-in-Bay on Lake Erie, where she developed a fascination for the freshwater Bryozoa. With the encouragement and supervision of the Laboratory Director, Dr. Raymond C. Osburn, she wrote her doctoral dissertation on the bryozoans of Lake Erie and the anatomy of *Lophopodella carteri*.

After receiving her Ph.D. degree in 1934, Mary Rogick joined the biology faculty at New Rochelle College, New Rochelle, New York, eventually rising to the rank of full professor and department chair. Despite a heavy and unremitting teaching load and a professional environment that discouraged research, Dr. Rogick remained highly active in her field. She published 43 papers concerning freshwater and marine Bryozoa, a laboratory manual in general zoology, nine articles about teaching methods, and several miscellaneous papers. Most of this work was done from the small second-floor apartment where she lived for 28 years.

An accomplished illustrator, Dr. Rogick's papers were distinguished by their clear and detailed drawings, one of which is reproduced here in Figure 15a. In addition, she was skilled at drawing simple cartoons, and often decorated her correspondence with humorous characters and a whimsical signature.

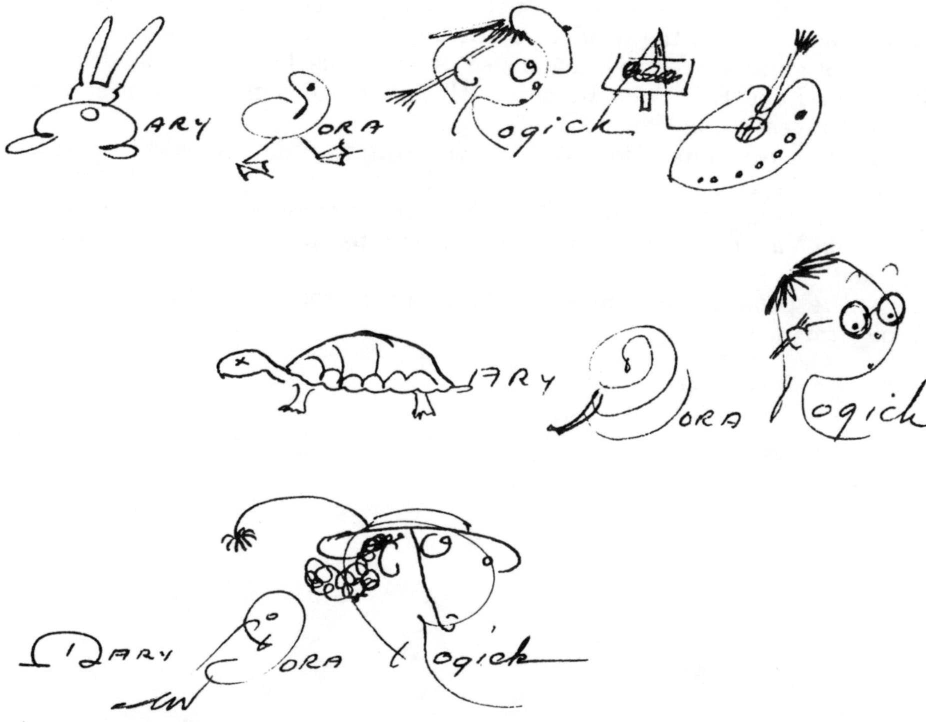

With humility, dedication, and perseverance, Mary Rogick set high standards in teaching and research. She touched many lives, inspiring students and colleagues alike to overcome apparent obstacles, and to achieve their very best.

(Photo courtesy of the Archives of New Rochelle College, New Rochelle, New York).

ACKNOWLEDGEMENTS

The state survey which formed the basis of this work benefitted from the enthusiastic and able assistance of several undergraduate and graduate students at Wright State University. Special thanks are extended to: Daniel L. Bowker, James M. Lazorchak, Kane J. More, Betty A. Orr, Georgie-Ray Rosier, Lois Ann Stoffel, Bruce M. Wahle, and Michael W. Zimmerman.

At parks and wildlife areas throughout Ohio, we received valuable assistance and advice from the resident staffs and supervisors. In the Lake Erie portion of the state survey, we made use of generous facilities provided by the Center for Lake Erie Area Research, College of Biological Sciences, The Ohio State University, at Put-in-Bay.

I would like to thank Veda M. Cafazzo, Editor, Ohio Biological Survey, for her editorial services, and Rosemary Kullman, Secretary, Ohio Biological Survey, for entering the manuscript on the word processor, and thus setting the type. Vicki Vaughan was instrumental in refining some of my rudimentary line drawings.

The entire project was funded through two consecutive grants from the Ohio Biological Survey.

CONTENTS

LIST OF FIGURES

LIST OF TABLES

PREFACE

Bryozoa are among the most common animals living on submerged objects in shallow fresh water. In Ohio, from May through October, an examination of aquatic habitats almost always reveals one or more species.

Viewing them using low-powered magnification, it is difficult not to become fascinated quickly by these animals. What appears to be an individual is actually a colony of zooids. Their whorls of delicate feeding tentacles swaying slowly in the water resemble tiny flowers. The assortment of small invertebrates which grow on and around many bryozoan colonies can be likened to a microscopic coral reef community in its abundance and diversity.

Despite being so common, freshwater bryozoans are poorly known. They are easy to overlook in the field because they resemble only inert scraps of brown moss or blobs of jelly. However, their reputation for being difficult to recognize, collect, grow, preserve, and identify is not entirely justified.

From a strictly biological standpoint, bryozoans offer many exciting opportunities for study. They are anatomically simple, easily cloned, and rapid-growing. Their life cycles include sexual and at least two forms of asexual reproduction, a true diapause, and a well-developed motile phase. Colony growth and development are still poorly understood: preliminary genetic studies have revealed a number of puzzles. The taxonomy needs much attention; and many fundamental aspects of sexual reproduction, digestion, and respiration are completely unknown. This guide was written to provide basic tools and information that may be useful in studies such as these.

Timothy S. Wood
Department of Biological Sciences
Wright State University
Dayton, OH 45435

BIOLOGY OF ECTOPROCT BRYOZOANS

Ectoproct bryozoans are small marine and freshwater animals which comprise the Phylum Ectoprocta. They are colonial organisms composed of many similar connecting zooids, each with its independent food gathering structure, mouth, digestive tract, muscles, nervous system, and reproductive ability. Despite their apparent autonomy, zooids share certain tissues and fluids which unify the colonies physiologically.

Of the three classes of ectoproct bryozoans, by far the best known is Gymnolaemata, with several thousand (mostly marine) species, and an extensive fossil record. In contrast, the Phylactolaemata has only about sixty known species, all extant and found exclusively in fresh water. The families belonging to this class have not been organized into orders. The discussion that follows focuses mainly on the Phylactolaemata.

BASIC ANATOMY

Among phylactolaemate Bryozoa, the individuality of zooids is blurred to the extent that it is impossible to distinguish precisely where one zooid stops and the adjacent one begins. An inventory of zooid structures includes the following (see also Figure 1):

Polypide. A complex of all major organ systems is suspended in an internal coelomic fluid which circulates to all parts of the colony. The polypide includes the following parts:

LOPHOPHORE: a food-gathering structure bearing many ciliated tentacles which may be either extended flower-like during feeding, or collapsed and completely withdrawn into the interior of the colony;

MOUTH: situated centrally at the base of the tentacles; phylactolaemates have a special lobe (epistome) which hangs over the mouth and which is believed to have an important sensory function;

GUT: the most prominent feature is a long caecum in which ingested particles are mixed thoroughly with vigorous peristaltic contractions;

FUNICULUS: a thin cord of tissue loosely joining the end of the gut to the colony wall. The funiculus is the site of statoblast production and spermatogenesis, and;

CENTRAL NERVE GANGLION: inconspicuously located between the mouth and the anus, with a major nerve tract extending into each arm of the lophophore.

Cystid. This is the laminated living and nonliving structure that separates the coelom from the external environment. Its outermost layer (ectocyst) consists of secreted material, which in some species is a slimy mucus, while in others it is a chitinous, somewhat leathery cuticle. Occasionally the cystid is called a zooecium (from the Greek, *zoos* = animal and *ecos* = shelter), a remnant of the time when this was believed to be a nonliving case within which the bryozoan animal lived.

In tubular, branching colonies the body is often encrusted with tiny particles from the surrounding water. A distinct ridge, or keel, may extend longitudinally along the side opposite the substrate (Figure 21a). If the peak of the ridge is free of encrusted particles, it is called a furrow.

The cystid also includes strands of circular muscle which help to regulate the internal pressure of the coelomic fluid. Tracts of beating cilia on the peritoneum circulate the fluid in random eddies among polypides throughout the colony.

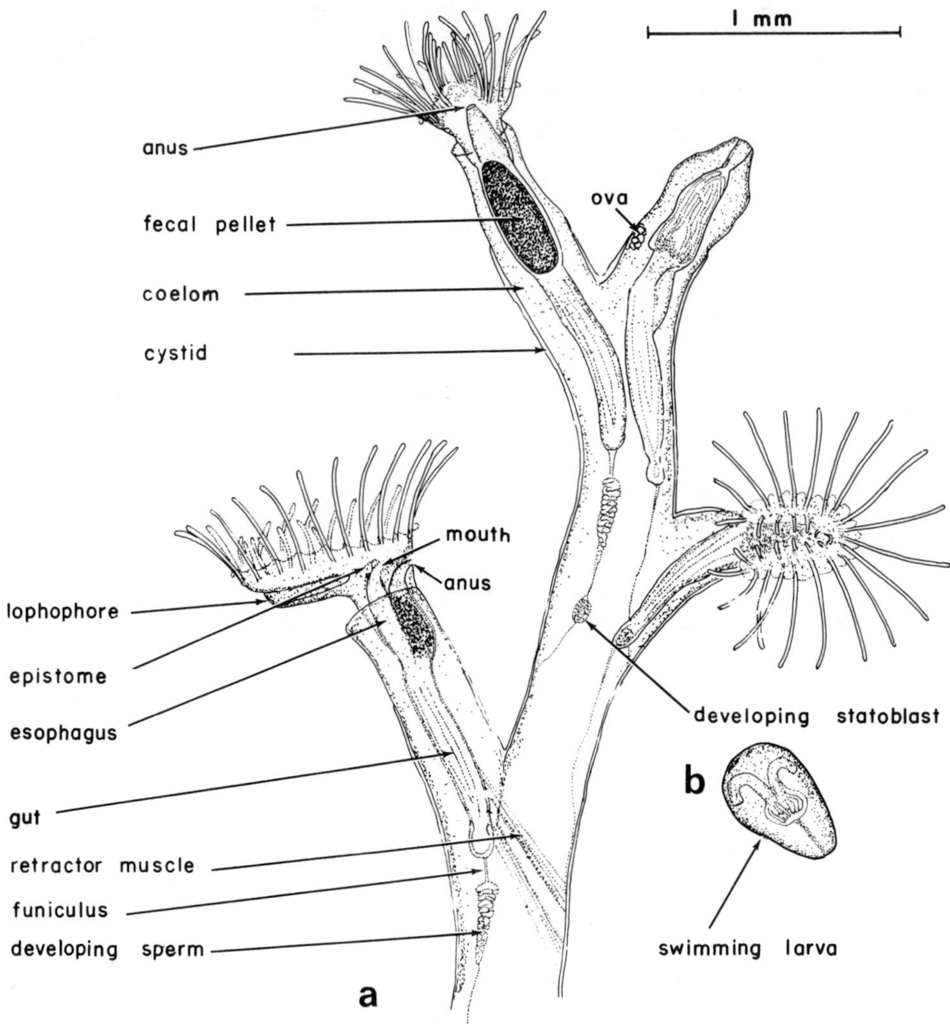

Figure 1. Generalized view of a freshwater bryozoan and a typical swimming larva. a-A plumatellid showing the location of major parts; b-larva showing two internal polypides, drawn to the same scale as Figure 1a. Larval morphology is fairly uniform for all species.

ASEXUAL REPRODUCTION

Asexual reproduction is well developed among freshwater bryozoans, and includes simple fragmentation and several types of budding.

Fragmentation occurs when a piece of a colony becomes separated from the rest and establishes itself elsewhere. In *Fredericella sultana* and *Plumatella fruticosa,* a free branch often breaks off, leaving behind a short stump. The branch, with several living zooids, drifts freely until it encounters a solid object to which the zooids adhere.

Colonies of *Lophopodella carteri* and *Cristatella mucedo* are similarly capable of fragmentation. These are slowly creeping colonies, portions of which separate from the main body and move off in different directions.

All phylactolaemate bryozoans form small, encapsulated structures called statoblasts. These develop on the funiculus from material derived from the colony wall. Most statoblasts undergo an obligate period of dormancy, and many can be dried or frozen for long periods. Under proper conditions they germinate, producing a single zooid which eventually may give rise to an entire new colony (Figures 2b-d).

Several general types of statoblasts are recognized on the basis of their morphology and function:

FLOATOBLASTS: have incorporated in their outer edge an extensive band of enlarged chambers filled with a gas which provides buoyancy (Figure 3a-c);

SESSOBLASTS: generally larger than floatoblasts, have no gas-filled chambers, and are cemented firmly to the substrate (Figure 3d-e), and;

PIPTOBLASTS: have no gas-filled chambers, and are not cemented to the substrate. They lie within the tubular colony interior, and may adhere to the colony wall by means of small keel-like projections (Figure 3f-g).

In many cases, statoblasts are essential for species identification. The outer casing of each statoblast can be separated into two opposing halves, or valves, which fit together like the two halves of a clamshell. Each valve has an inner and an outer layer (Figure 4). The inner layers together form a central capsule, which encloses germinal tissue and yolky materials. The outer layer, or periblast, varies according to species, and sometimes bears ornamentation. In a floatoblast, it is the periblast that contains the band of buoyant cells. This band, or annulus, is wider on one valve than on the other, and it encloses a clear, central area known as the fenestra. Traditionally, that valve with the widest annulus is designated the dorsal valve, while its opposite member is the ventral valve.

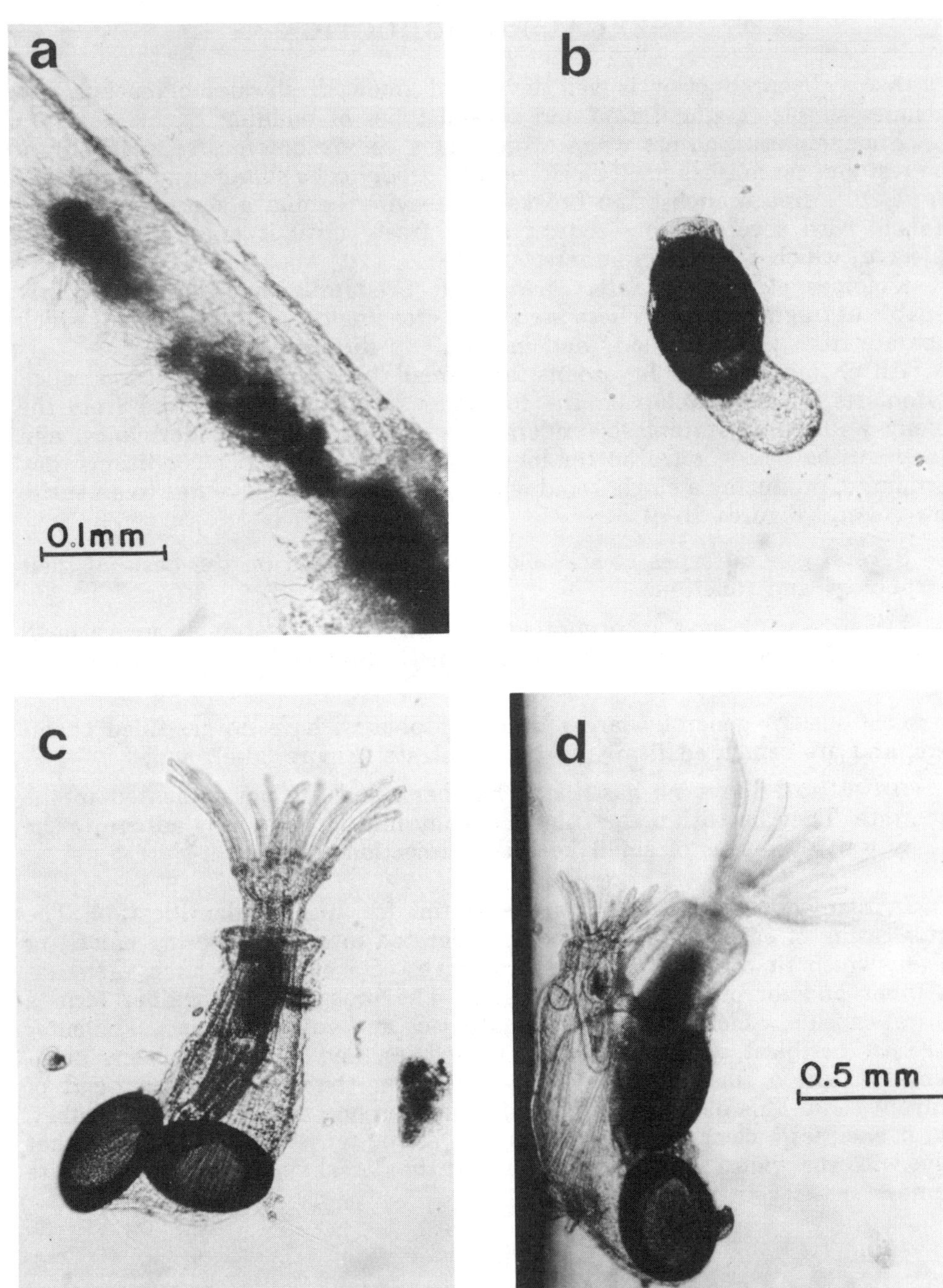

Figure 2. Developmental stages of bryozoans. a-developing spermatozoa on a funiculus in *Fredericella indica;* b-extension of an adhesive pad from a germinating statoblast of a new colony in *Plumatella casmiana;* c-growth of the first zooid (ancestrula) from the germinated statoblast of *P. casmiana;* d-appearance of the second zooid in the new *P. casmiana* colony.

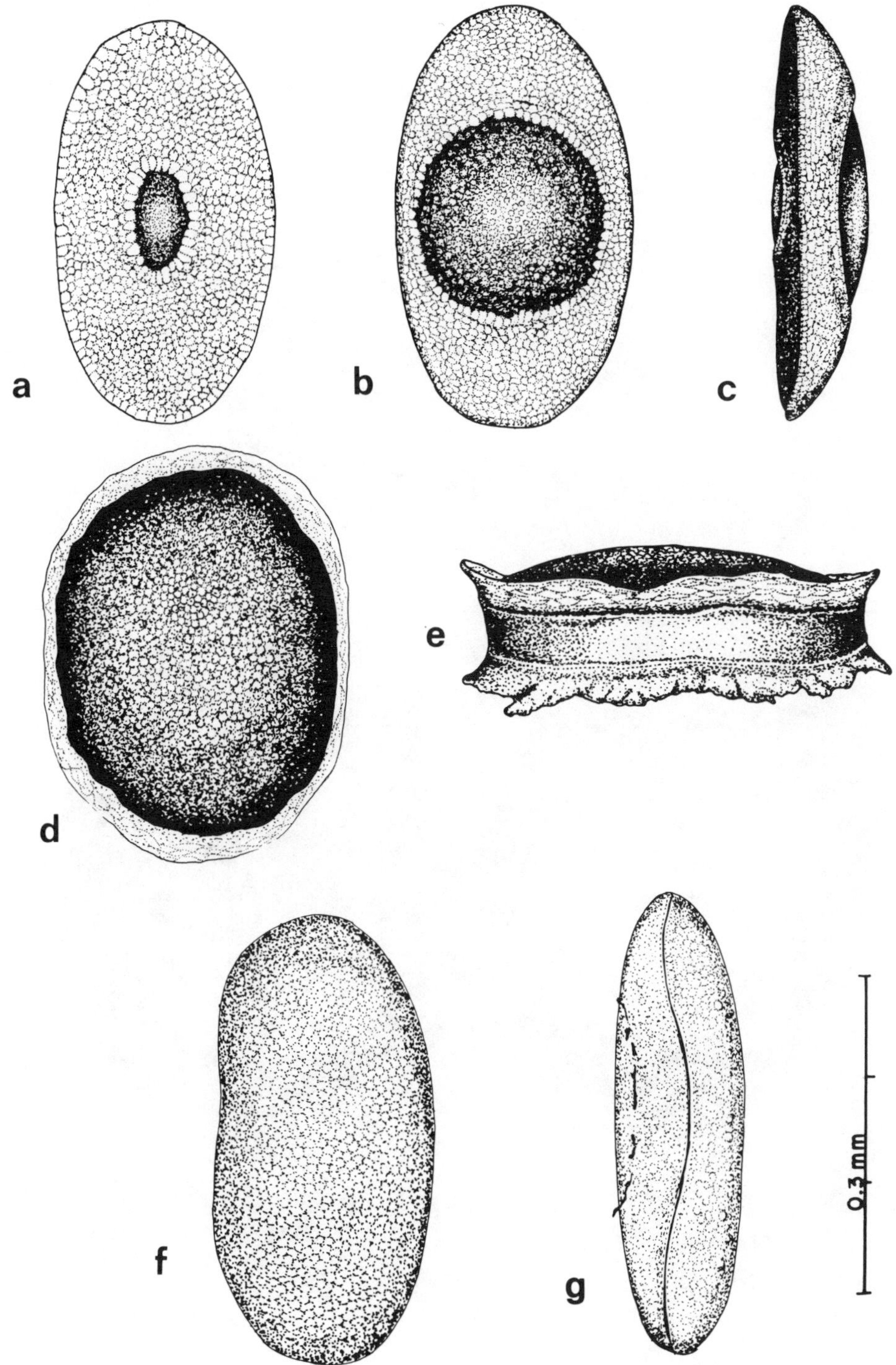

Figure 3. Three basic types of statoblasts. a-c are dorsal, ventral, and lateral views of a floatoblast of *Plumatella emarginata;* d-e are frontal and lateral views of a sessoblast of *P. emarginata,* and; f-g are dorsal and lateral views of a piptoblast of *Fredericella indica.*

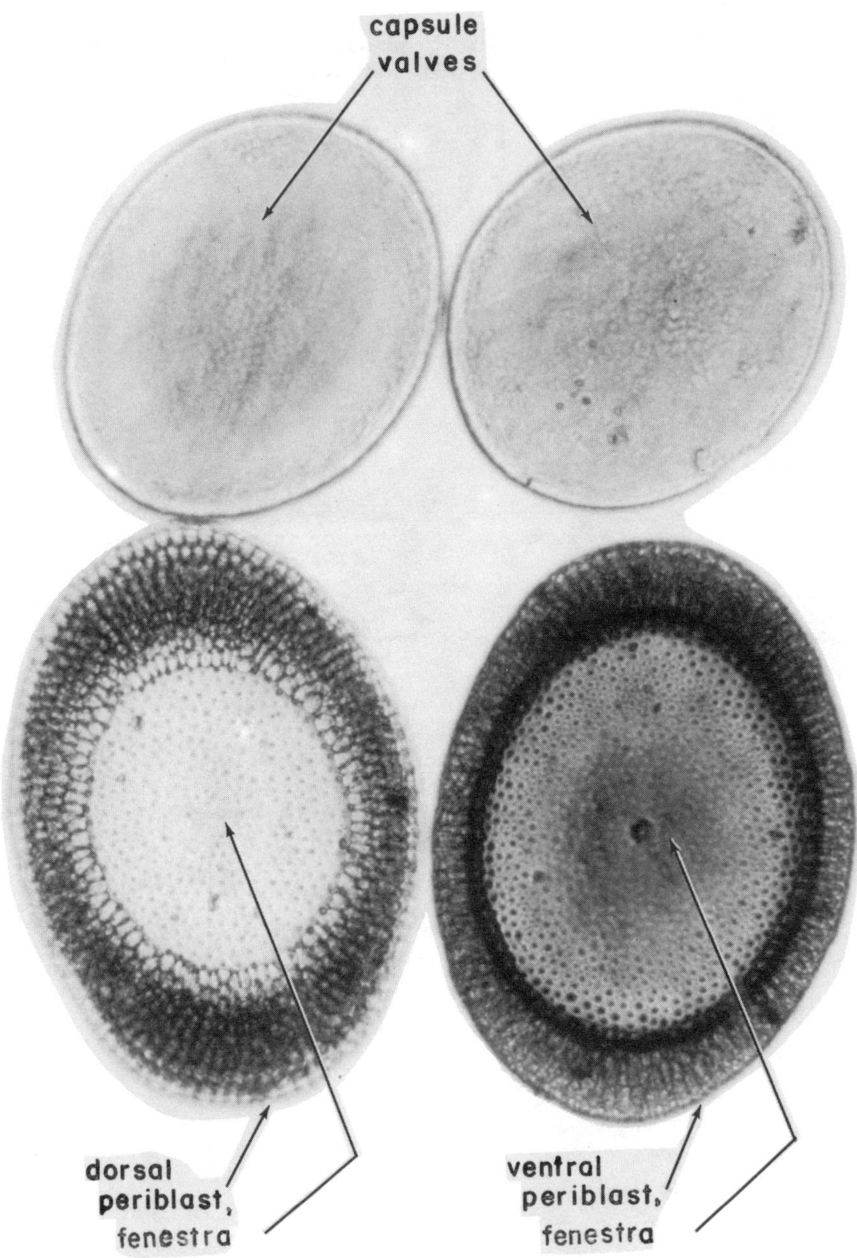

Figure 4. Dorsal and ventral valves of *Plumatella repens* floatoblast separated into composite parts.

SEXUAL REPRODUCTION

Sexual reproduction in the freshwater Bryozoa normally occurs in Ohio for a few weeks during the late spring or summer. Even among laboratory-reared colonies, sexual activity occurs mainly at those times when one would expect it in natural populations.

A single colony produces both eggs and sperm. The sperm develop in clusters along the funiculi of polypides in certain regions of the colony (Figures 1a; 2a). Upon release into the coelom, large numbers circulate passively, resembling inert, kinked threads. Eggs appear as tiny grape-like clusters of 20-40 cells attached to the inner colony wall ventral to the polypide and near the next developing bud (Figure 1a).

Although no exchange of gametes has been observed among colonies, there is a strong likelihood of cross-fertilization. In certain species, the zooids emerging from independently germinating statoblasts fuse, forming a single colony with common coelom (Mukai et al, 1984). The union of gametes from those portions of the colony having separate origins probably results in a true mixing of hereditary material. In those colonies arising from a single statoblast, self-fertilization is the rule.

In *Lophopus crystallinus* and possibly other species, only a very few eggs develop fully to ciliated larvae (Marcus, 1934). This development occurs in a special embryo sac, where the larva grows to fill the space normally occupied by a polypide. The so-called larva (actually a specialized motile phase) eventually assumes a wedge shape and is released to the surrounding water through a terminal pore. It contains one or more fully-developed polypides enclosed within a ciliated mantle (Figure 1b). The larva may swim from several minutes to more than 24 hours, probing various substrates. Substrate discrimination by larvae has been demonstrated by Hubschman (1970) in one species, *Pectinatella magnifica*. Within an hour after settling, the mantle pulls back, the tiny lophophores are extended, and feeding begins.

STUDY METHODS

COLLECTING BRYOZOANS

In Ohio, the best places to find bryozoans are lakes, ponds, and streams where there is plenty of suitable substrate. Most colonies occur on the sides and undersides of submerged objects, such as floating or submerged wood, aquatic plants, and rocks. The shallow bays and upper regions of impoundments where old substrates have accumulated often are excellent collecting sites. Rocky streams can be good collecting sites if not subject to frequent flooding. Generally, poor sites for bryozoans include wave-swept shores and soft-bottomed lakes lacking solid substrate.

Colonies can be collected best by removing them together with a portion of the material on which they are growing. In most species, scraping a colony from its substrate causes heavy damage, and usually produces only fragments.

Collecting equipment should include:

a short-bladed knife for removing wood or other soft substrate on which a specimen is growing. The smallest available hunting knife works well;

a good 10x or 14x hand lens, worn on a lanyard around the neck. The single-lens Coddington type is preferred, because other varieties become useless when water enters between lens components if the lens is accidently submerged, and;

a cold chisel, hammer, and safety glasses for collecting colonies from rock substrates. Chipping pieces of rock with attached colonies is best done on a solid surface, such as pavement or a large, fixed rock. The shock of hammering generally does little damage to the bryozoan.

Most species can be collected in shallow water while the collector is wading or snorkeling. Buoyant statoblasts tend to adhere to floating pieces of styrofoam, and it is useful to check such trash along the shore with a hand lens to see what bryozoan species may be present in the area.

PRESERVATION AND STUDY

Once removed from their natural habitat, nearly all species can be maintained for many days in gently aerated water. One exception is the large colony of *Pectinatella magnifica,* which rapidly disintegrates when confined.

To preserve bryozoans with the lophophore extended, it is important to narcotize the specimens before fixing them. The least troublesome and most consistently satisfactory narcotizing agent is menthol. In a small dish of water, in which the specimen has its lophophores fully extended, menthol crystals are sprinkled over the surface. The dish is then covered and allowed to remain undisturbed for 1-3 hours. The period of narcotizing varies with temperature and species (*Lophopodella* requires less than the average time, *Fredericella* needs considerably more). If left too long, the lophophores will begin to disintegrate. Well narcotized specimens appear fresh and immobile, are insensitive to gentle prodding, and do not draw a current of water between the tentacles. At this point they may be transferred directly to a fixative.

Less messy than menthol crystals are customized menthol wafers. To make these, shape pieces of aluminum foil around coins to form shallow molds. Then, melt menthol crystals over low heat and pour the liquid into the molds. Cool until solid, remove the foil, and store the wafers in an air-tight container.

Any of the standard invertebrate fixatives may be used with bryozoans. A 10% formalin solution serves as an excellent fixative, and can be diluted to 5% for long-term preservation. Seventy percent ethanol also can be used, but its volatility makes tiny specimens difficult to manipulate under a microscope. Bouin's fixative is preferred for most histological work, although formalin is still best with statoblasts.

Because statoblasts play such an important role in species identification, it is useful to maintain a separate collection or record of those taken with each colony specimen. To divide statoblasts into their component parts, Mukai and Oda (1980a) immersed them in 10M KOH for three hours at room temperature. Wiebach (1974) preferred the faster method of heating them for three minutes with 2.5M KOH in a small nickel spoon over an open flame. In either case, the statoblasts are treated individually to avoid mixing parts. The capsule and periblast valves either separate spontaneously or can be gently teased apart after treatment. The organic contents usually emerge cleanly as a single mass without adhering to the capsule.

After being washed in water, statoblasts can be dehydrated in ethanol and mounted in balsam or various synthetic media. However, for individual statoblasts, a photographic record is more useful and easier to prepare (Figure 4). For this purpose, the statoblast parts are transferred to a drop of aged water on a microscope cavity slide and arranged in a suitable position for photography. An ordinary compound microscope with camera back and substage lighting is satisfactory.

HISTOLOGICAL PREPARATION

For light transmission microscopy, Mukai and Oda (1980a) achieved excellent results by embedding tissue in paraffin, staining with Delafield's hematoxylin and counter-staining with eosin, then sectioning at 7 μm. Davenport (1890) used Czoker's cochineal to stain embryonic cells of the bud in *Cristatella.*

For chromosome preparations, Backus and Wood (1971) immersed dissected tissue in 0.05% w/v colchicine in tapwater for no more than 55 minutes, and then fixed the material in a solution of 32% acetic acid, 32% methanol, and 36% ethanol. A small amount of tissue was then placed on a microscope slide and treated with acetic acid, washed in distilled water, and stained in a 3% dilution of stock Giemsa in 0.01M phosphate buffer at pH 7. After being washed in distilled water, slides were warmed (65°-70° C) for at least ten minutes, placed in xylene for ten minutes, air-dried, and coverslip mounted in a synthetic resin.

Potter (1979) successfully used lactic-acetic orcein stain for chromosomes of *Plumatella, Cristatella,* and *Pectinatella,* then destained with lactic-acetic acid for *Pectinatella.*

For transmission electron microscopy, Franzén (1982) used the following procedure: he immersed tissue in 5% gluteraldehyde in either 0.2M cacodylate or phosphate buffer at 5°C. After fixation, he washed the material in cold buffer and postfixed it in cold 1% osmium tetroxide using the same buffer. He dehydrated with ethanol and processed through polypropylene oxide into Epon. Finally, he sectioned at 500 Å and stained with uranyl acetate for 30 minutes followed by lead citrate for 3 minutes.

Scanning electron microscopy requires the same pre- and postfixation. Material is then dehydrated in an ethanol series, transferred to filtered Freon TF, and critical-point dried from carbon dioxide (Franzén and Sensenbaugh, 1983). With statoblasts, the procedure is simpler. Mundy (1980) cleaned statoblasts with detergent for a few minutes in an ultrasonic bath, washed them in distilled water, and dehydrated them through acetone over two hours. Air drying gave satisfactory results, with only occasional distortion. The statoblasts were sectioned with a razor blade, then coated with gold.

GERMINATING STATOBLASTS

Nearly all statoblasts are dormant for variable lengths of time, from several weeks to many months. When left in water at room temperature they eventually germinate, one by one. However, reliable and synchronous germination usually can be induced by storing the statoblasts for at least several weeks in cold or dry conditions, then placing them in water at room temperature.

Statoblasts of *Plumatella* and *Fredericella* species can be stored in a refrigerator, either dry or in water. Because these are so small, it is convenient to store entire colonies with the statoblasts inside, then to dissect out the statoblasts as needed. Sometimes the percent germination of *Fredericella* statoblasts can be increased by drying them for at least one week prior to use.

Statoblasts of *Cristatella* and *Pectinatella,* when stored in water in a refrigerator, collect in dense clumps on which there often develop fatal growths of fungi. *Pectinatella* statoblasts do not survive long under dry conditions. Therefore, both species can be stored most successfully in ice. First, the statoblasts should be held for several days in cold water. Then, pea-sized clumps of the statoblasts should be frozen in ice cube trays with glass cover slips holding the clumps below the surface while the water is freezing. Store the trays in the freezer in sealed plastic bags. Germination of this material is most successful when it is done under slight environmental pressure, such as beneath several centimeters of water.

Lophopodella tolerates dry conditions extremely well, especially at cool temperatures, and may be refrigerated either wet or dry. Entire colonies filled with statoblasts may be dried and kept for years in a refrigerator or freezer. Upon returning to water at room temperature, the statoblasts germinate almost simultaneously within four or five days. These too benefit from environmental pressure to encourage germination.

OBTAINING LARVAE

Colonies suspected of containing larvae should be placed in a dish of water and kept undisturbed overnight. In the laboratory, as in the field, larvae normally are released at between the hours of 11 PM and 3 AM. However, *Pectinatella* frees larvae whenever the colony is disturbed.

Resembling miniature blimps in the water, swimming larvae are easily visible to the unaided eye. They can be transferred to other containers by using an eyedropper pipette. Settling may occur within minutes or up to 24 hours after release. Larvae settle most readily on substrates that have been immersed in fresh pond water for at least 18 hours.

REARING COLONIES IN THE LABORATORY

Bryozoan colonies in the laboratory must be grown upside down to avoid fouling by their own fecal pellets and other settling debris. Petri dishes (or their covers) make excellent substrates, because these can be turned right-side-up for extended microscopic examination, and the colonies may be viewed with transmitted light, if desired (Figure 5b).

When starting colonies from nonbuoyant statoblasts, a simple procedure is to allow them to germinate in a covered dish of water. Several days of growth usually are required before the individual becomes so firmly attached that the dish can be inverted in a larger container.

In the case of floatoblasts, the petri dish must be inverted in a tray of water, with no bubbles trapped inside. When statoblasts are gently squirted under the inverted dish by means of a pipette, they will rise and come to rest against the inside of the dish. Attachment follows shortly after germination.

Scraps of a colony sometimes can be transferred to a petri dish for rearing in the laboratory. *Lophopodella, Cristatella,* and small colonies of *Pectinatella* are simply nudged from their natural substrate and gently transferred to the dish. They sink and become attached to the petri dish within a day or so. Free branches of *Fredericella* and *Plumatella* colonies may be treated in a similar manner, but these usually require more time, and attachment is made only at the tips of young zooids and any subsequent growth.

Bryozoans derive nourishment from the suspended detritus which occurs naturally in a large, thriving aquarium. A simple, two-tank system is most effective. Inverted petri dishes with their attached colonies are kept on racks in a small glass tank which may be built specifically for this purpose (Figure

5a). The racks should be tilted so that any trapped bubbles will accumulate at the higher end of the dishes. Water is circulated through the tank from a larger aquarium in which most of the organic food is generated. The presence of well-fed goldfish in the larger tank creates a favorable environment for most byrozoan species, probably because the fish keep organic particles in suspension. The water must not be filtered, and a portion of it should be changed every few weeks. If available, a pond or lake is the preferred source of water. Organic sediment should be allowed to accumulate at the bottom of both tanks. The smaller rearing tank may be darkened to limit algal growth.

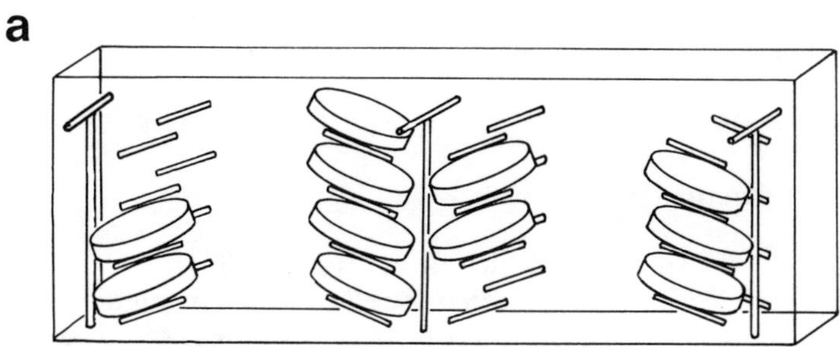

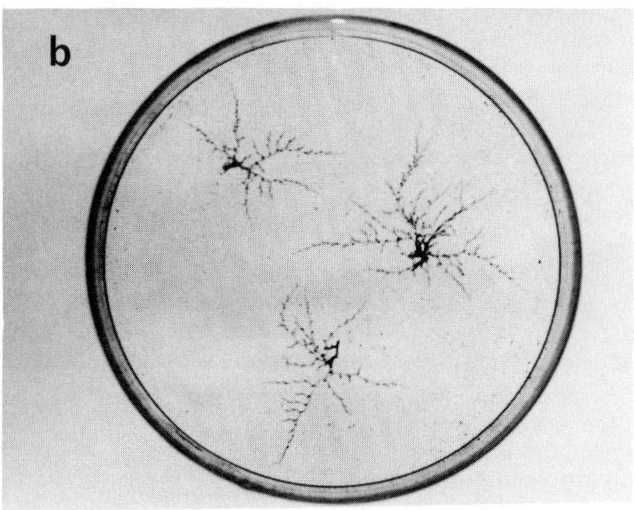

Figure 5. Laboratory rearing of bryozoans. a-Glass tank used for laboratory rearing of bryozoan colonies. Four-inch (10.2 cm) diameter petri dishes rest inverted on solid glass rods cemented to the inside walls. Water is circulated continuously from a larger aquarium housing active, well-fed fish. External dimensions of the rearing tank are: 66 cm long, 23.5 cm high, and 11.5 cm wide; b-three colonies of *Plumatella repens* growing in the laboratory on a glass petri dish from a rearing tank.

Occasionally, bryozoan colonies in the laboratory stop growing. Zooid longevity is shortened and the budding rate drops to near zero. Although this problem apparently is caused by conditions in the water, often not all colonies are affected equally. Replacing some or all of the water with fresh lake or pond water eventually can stimulate renewed growth. Feeding with "infusorial powder" or frozen food suspension (used by marine aquarists) also may help revive senescent colonies.

CHECKLIST OF FRESHWATER BRYOZOAN SPECIES IN NORTH AMERICA

The following species of freshwater Bryozoa have been reported from North America. An asterisk indicates those species which have been collected in Ohio. The sequence of taxa within this checklist does not imply phylogeny.

Phylum Ectoprocta

    Class Phylactolaemata Allman, 1856

        Family Fredericellidae Hyatt, 1868
* *Fredericella indica* Annandale, 1909
* *Fredericella australiensis* Goddard, 1909

        Family Plumatellidae Allman, 1856
* *Plumatella casmiana* Oka, 1907
* *Plumatella repens* (Linneaus, 1758)
* *Plumatella fungosa* (Pallas, 1768)
* *Plumatella emarginata* Allman, 1844
* *Plumatella reticulata* Wood, 1988
* *Hyalinella punctata* (Hancock, 1850)
  Hyalinella orbisperma (Kellicott, 1882)
  Plumatella fruticosa Allman, 1884
  Plumatella coralloides Annandale, 1911
  Stephanella hina Oka, 1908
  Stollela indica Annandale, 1909
  Stollela evelinae Marcus, 1941

        Family Lophopodidae Rogick, 1935
* *Lophopodella carteri* (Hyatt, 1866)
* *Pectinatella magnifica* (Leidy, 1851)
  Lophopus crystallinus (Pallas, 1768)

        Family Cristatellidae Allman, 1856
* *Cristatella mucedo* Cuvier, 1798

    Class Gymnolaemata Allman, 1856

      Order Ctenostomata Busk, 1852

        Family Paludicellidae Allman, 1856
* *Paludicella articulata* (Ehrenberg, 1831)
* *Pottsiella erecta* (Potts, 1884)

        Family Victorellidae Hincks, 1880
  Victorella pavida Saville Kent, 1870

FRESHWATER HABITATS IN OHIO

The State of Ohio covers an area of 106,764 km², including over 680 km² of inland waters, but excluding 8,954 km² of Lake Erie (Figure 6). There are roughly 71,000 km of named streams and rivers, and over 60,000 lakes, ponds, and reservoirs. The state has a squarish outline and lies mostly between 39 and 41 degrees north latitude. The Ohio River flows 724 km along its southern border, and in the north is 492 km of Lake Erie shoreline.

Few of Ohio's lakes and ponds are natural. Most of these are farm ponds under one hectare in size. Others are reservoirs constructed for purposes of flood control, low flow stream augmentation, or as feeder reservoirs to the

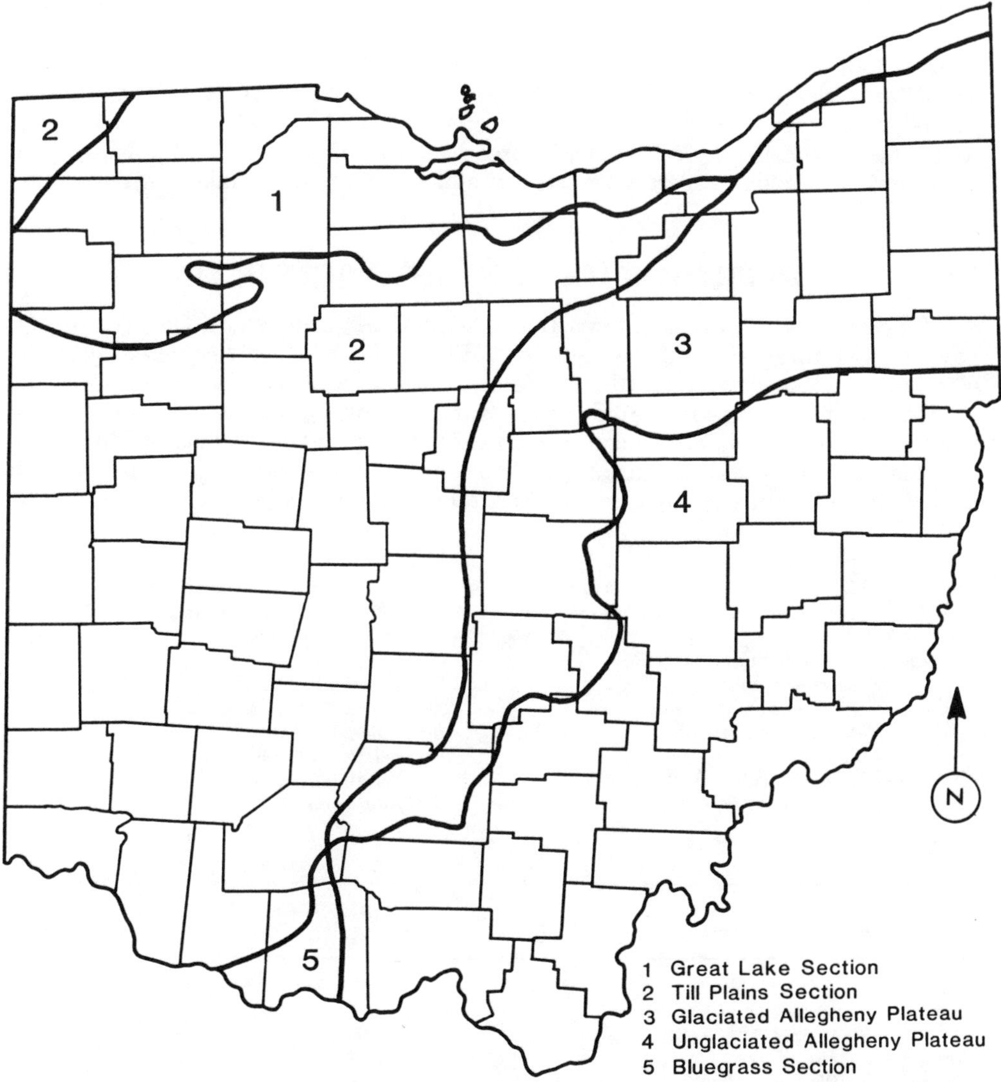

Figure 6. Physiographic sections of Ohio. Modified from Anderson (1983).

state's two canal systems (long since abandoned). Ohio has 724 km of rivers which are recognized specifically in state or national Scenic Rivers programs. Unfortunately, large sections of other streams have been badly disfigured by channelizing for drainage and flood control.

In the following account, the names and descriptions of Ohio's physiographic sections are based upon Anderson (1983). Most of western and north-central Ohio lies in a zone of fertile land known as the Till Plains Section. These flat to gently rolling plains were shaped by a succession of ice sheets in the very recent geological past. Lakes in the area are naturally eutrophic, with hard, somewhat alkaline water. Where substrate is available the lakes support a diverse community of bryozoans.

Extending across the far northern part of the state is the flat Great Lake Section. Rivers here are sluggish and muddy, and the widely scattered lakes are small. Most sites with public access lack substrate suitable for bryozoans, so in this region few colonies have been found.

The Glaciated and Unglaciated Allegheny Plateaus cover most of the eastern half of Ohio. Lakes in the northern third are similar to those in the Till Plains Section, though possibly less productive. Farther south, in the rugged, unglaciated portions of the state, soils are thin and lakes are usually somewhat acidic. Many of the widespread coal mining areas suffer from acid mine drainage, which severely restricts the biotic diversity of lakes and streams.

DISTRIBUTION OF OHIO BRYOZOA

The bryozoan species known to occur in Ohio, and where they currently have been found, are summarized in Table 1. This list is restricted to sites offering public access, including most state parks and state wildlife areas.

Many of the sites in Table 1 were surveyed only once, which is probably insufficient to establish an accurate inventory. Populations fluctuate throughout the season as well as annually, and some species surely were missed. It is hoped that future collections will correct any omissions.

Table 1. Public access sites in Ohio where bryozoan species are known to occur; indicated by a +. Following the site name, one asterisk * represents a State Park, ** represent a State Wildlife Area, and *** represent a State River Access. The species are numbered from left to right according to the following code:

1=*Fredericella indica*
2=*Fredericella australiensis*
3=*Plumatella casmiana*
4=*Plumatella repens*
5=*Plumatella fungosa*
6=*Plumatella emarginata*
7=*Plumatella reticulata*
8=*Hyalinella punctata*
9=*Lophopodella carteri*
10=*Pectinatella magnifica*
11=*Cristatella mucedo*
12=*Paludicella articulata*
13=*Pottsiella erecta*

COUNTY	SITE	SPECIES CODES:												
		1	2	3	4	5	6	7	8	9	10	11	12	13
Adams	Adams Lake *	+			+									
Ashtabula	Orwell Wildlife Area **	+			+									
Athens	Dow Lake *	+												
Belmont	Piedmont Lake *			+	+									
Brown	Grant Lake **			+		+	+	+						

Table 1.—Continued

COUNTY	SITE	SPECIES CODES:												
		1	2	3	4	5	6	7	8	9	10	11	12	13
Carroll	Atwood Reservoir **			+	+				+		+			
Champaign	Kiser Lake *			+	+				+					
Clark	Clarence Brown Reservoir *			+			+	+			+			
Clermont	Stonelick Lake *					+	+				+			
Clermont	Harsha Lake *							+	+		+			
Clinton	Cowan Lake *	+	+	+	+		+	+	+		+			
Columbiana	Highlandtown Wildlife Area **				+		+				+			
Coshocton	Wills Creek Reservoir **	+						+						
Cuyahoga	Rocky River Reservation					+	+						+	+
Defiance	Oxbow Lake **	+			+									
Delaware	Delaware Lake *				+		+		+					
Delaware	Hoover Reservoir										+			
Erie	Kelleys Island				+									
Franklin	O'Shaugnessy Reservoir			+			+	+						
Fulton	Harrison Lake *	+		+			+							
Gallia	Tycoon Lake **				+						+			
Geauga	Punderson Lake *	+										+		
Greene	Huffman Lake	+		+		+		+			+			
Guernsey	Salt Fork Lake *			+			+				+			
Hamilton	Sharon Lake			+	+		+							
Hancock	Van Buren Lake *			+										
Harrison	Tappan Lake			+				+	+					
Henry	Florida Access, Maumee River ***			+	+		+							
Highland	Rocky Fork Lake *							+			+			
Hocking	Lake Logan *			+	+		+							
Jackson	Jackson Lake *	+		+							+			
Jefferson	Jefferson Lake *	+			+								+	
Knox	Knox Lake **	+		+	+		+	+			+			
Lake	Blueberry Pond, Holden Arboretum				+									
Lawrence	Vesuvius Lake **	+												
Logan	Indian Lake *			+				+			+			
Lorain	Findley Lake *	+		+	+							+		
Madison	Madison Lake *						+							
Mahoning	Mill Creek Park												+	
Meigs	Forked Run Lake *										+			
Mercer-Auglaize	Lake St. Marys *			+		+	+	+						
Miami	Decker Lake	+			+		+	+	+					
Monroe	Monroe Lake **	+		+	+									
Montgomery	Englewood Reserve			+										
Morgan	Burr Oak Lake *										+			
Morrow	Mt. Gilead State Park *	+												
Muskingum	Dillon Reservoir *			+		+					+			
Noble	Seneca Reservoir			+					+		+			
Ottawa	Lake Erie (East Harbor) *	+		+	+		+	+	+	+	+	+	+	+
Paulding	Flatrock Creek Access ***			+	+		+							
Perry-Licking	Buckeye Lake *			+			+		+		+			
Pike	Lake White *	+			+						+			

COUNTY	SITE	SPECIES CODES:												
		1	2	3	4	5	6	7	8	9	10	11	12	13
Portage	West Branch Reservoir *	+			+		+				+			+
Preble-Butler	Acton Lake *	+	+	+	+		+	+	+		+			
Scioto	Roosevelt Lake *										+			
Scioto	Turkey Creek Lake *					+								
Shelby	Lake Loramie *				+		+		+		+			
Summit	Portage Lakes *	+		+			+		+		+		+	
Trumbull	Mosquito Creek Reservoir *						+	+	+		+			
Warren	Caesar Creek Lake *		+					+	+		+			

Figure 7. Map of the Counties of the State of Ohio.

KEY TO FRESHWATER BRYOZOA
OF THE GREAT LAKES REGION

1. Colony composed of tubular parts, zooids in linear series or crowded upright to form a dense mat; statoblasts without projections extending beyond margins; tentacles fewer than 65 (Figures 8a-f, 9a-b, 10d-j)
....2

1. Colony sac-like and gelatinous; statoblasts with spines or hooks projecting beyond margins; tentacles more than 65 (Figures 8g-h, 9c, 10a-c)
....15

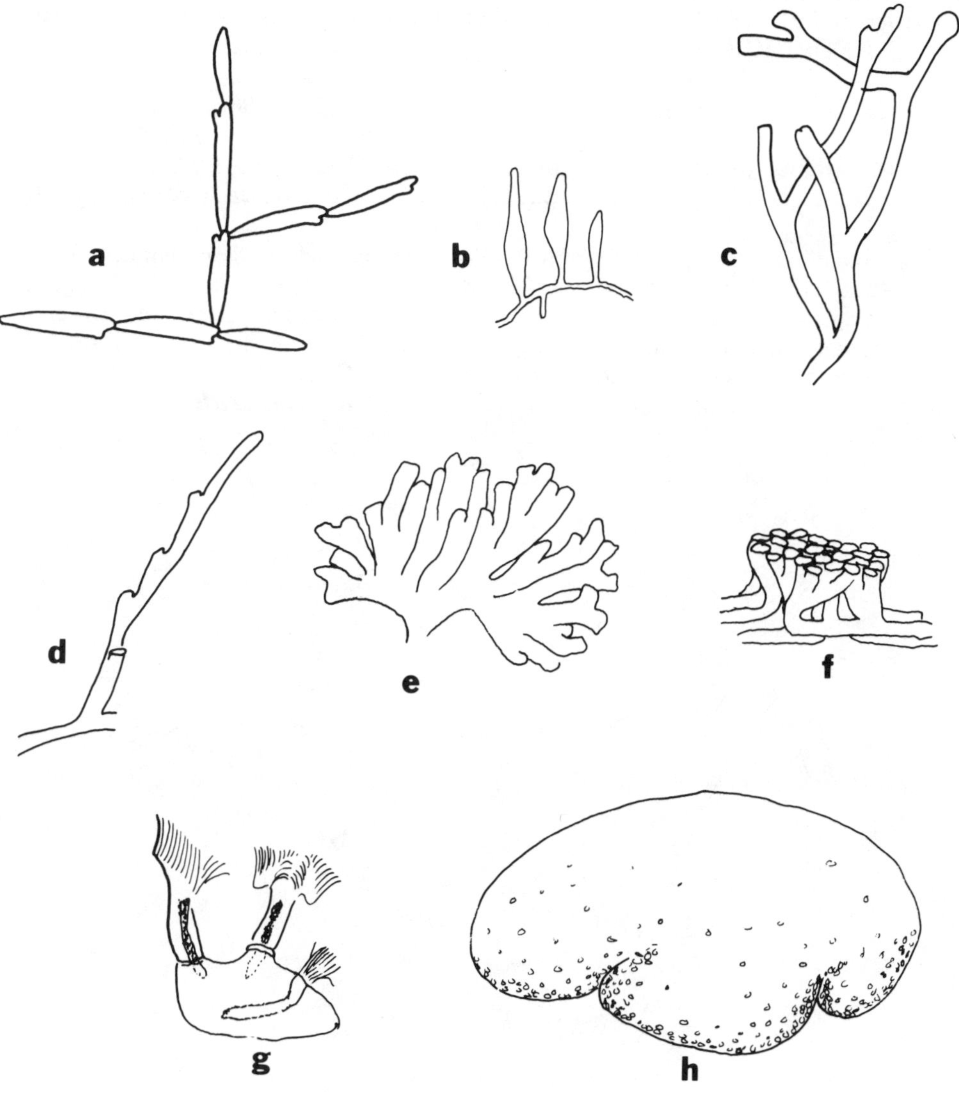

Figure 8. Colony structure in freshwater bryozoans. a—*Paludicella articulata*, X18; b—*Pottsiella erecta*, X18; c—*Fredericella indica*, X10; d— *Plumatella fruticosa*, X10; e—*Plumatella casmiana* on unrestricted substrate, X10; f—*Plumatella fungosa* or *Plumatella casmiana* on restricted substrate, X10; g—*Lophopodella carteri*, X12; h—*Pectinatella magnifica*, X1/5.

2. Extended lophophore circular in outline (Figure 9a); orifice round or quadrangular; statoblasts lacking annulus and never cemented to the substrate 3
2. Extended lophophore U-shaped (Figure 9b), orifice never quadrangular; statoblasts with extensive annulus or else cemented to the substrate. PLUMATELLIDAE 6

3. Orifice 4- or 5-sided; epistome and statoblasts absent; ectocyst stiff, shiny, and transparent; individual zooids clearly demarcated. .... PALUDICELLIDAE 4
3. Orifice round; statoblasts present or absent; ectocyst and zooids otherwise. FREDERICELLIDAE..5

4. Zooids branching from each other at various angles (Figure 8a)................. ....*Paludicella articulata* (p. 55).
4. Zooids arising upright from stolon-like parts (Figure 8b). ....*Pottsiella erecta* (p. 58).

5. Statoblast surface appearing dull and granular when dry; usually fewer than two statoblasts per zooid; common in Ohio.................................... ....*Fredericella indica* (p. 21).
5. Statoblast surface appearing shiny and smooth when dry; usually two or more statoblasts per zooid; uncommon in Ohio. ....*Fredericella australiensis* (p. 24).

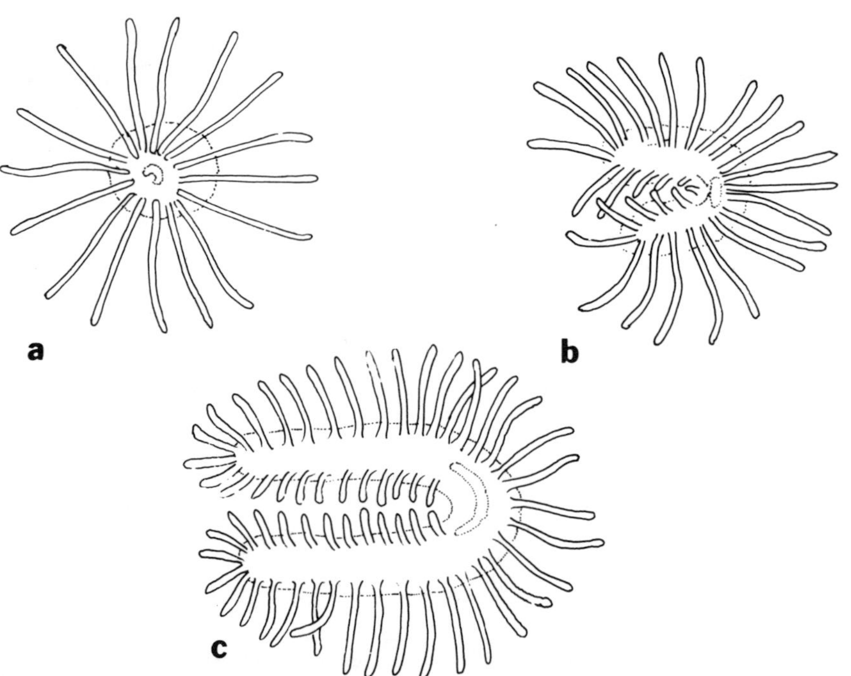

Figure 9. Lophophore variations, X15. a—*Fredericella indica;* b—*Plumatella emarginata;* c—*Cristatella mucedo.*

6. Zooids clustered in groups of 2 to 5, clusters joined by stolon-like cystids; colony wall delicate and transparent; unknown in Ohio......................14
6. Zooids not arranged in clusters; colony wall transparent to opaque.........7

7. Colony branches long and stringy, largely free of the substrate; statoblast length at least twice width; unknown in Ohio....................................
....***Plumatella fruticosa.***
7. Colony and statoblast otherwise...8

8. Colony wall clear and gelatinous; zooecial tips protuding slightly or not at all; statoblasts never cemented to substrate......................................9
8. Colony wall not gelatinous 10

9. Floatoblasts oval (Figure 10j). ....***Hyalinella punctata*** (p. 43).
9. Floatoblasts round; unknown in Ohio.................***Hyalinella orbisperma.***

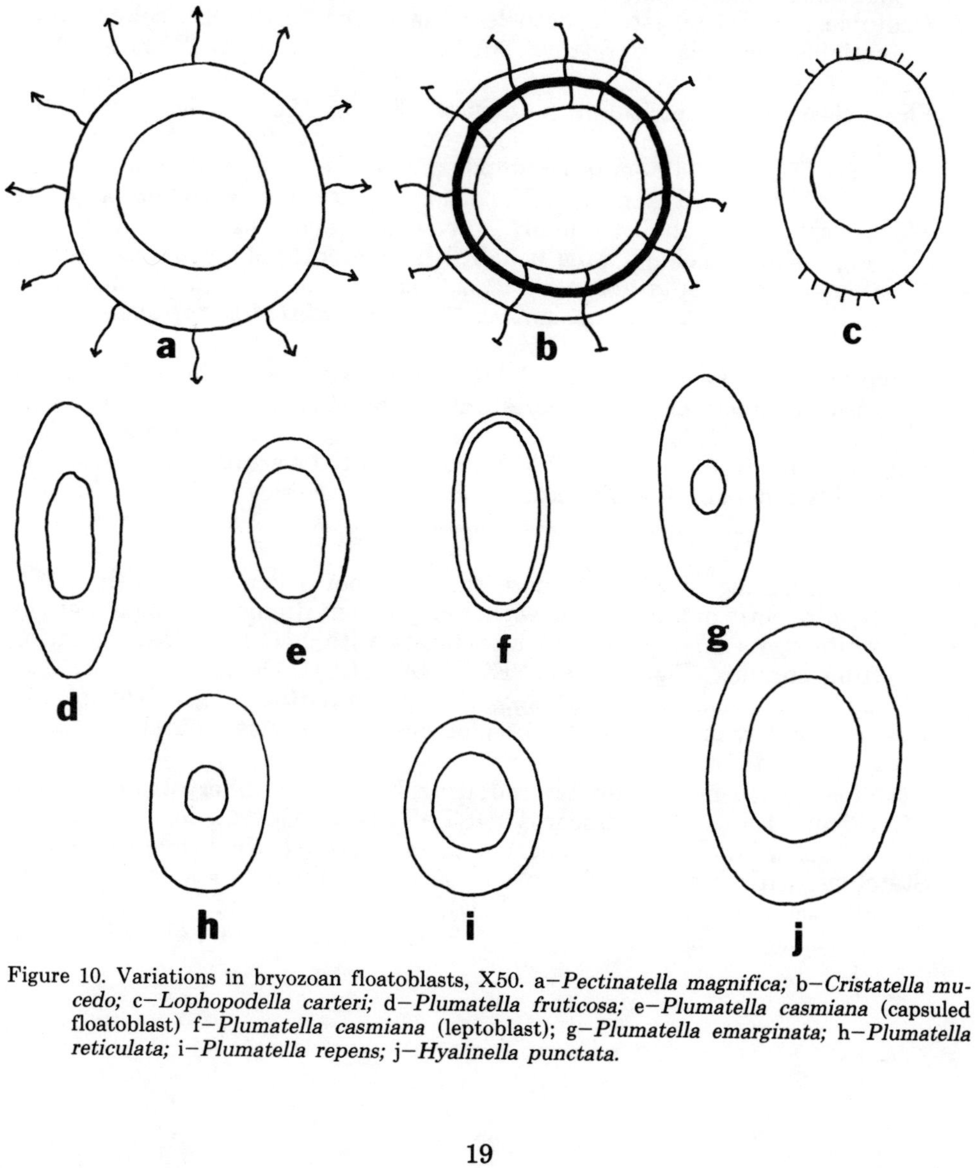

Figure 10. Variations in bryozoan floatoblasts, X50. a—*Pectinatella magnifica;* b—*Cristatella mucedo;* c—*Lophopodella carteri;* d—*Plumatella fruticosa;* e—*Plumatella casmiana* (capsuled floatoblast) f—*Plumatella casmiana* (leptoblast); g—*Plumatella emarginata;* h—*Plumatella reticulata;* i—*Plumatella repens;* j—*Hyalinella punctata.*

10. Dorsal annulus of floatoblast is extensive, maximum dorsal width of annulus equal to or greater than length of dorsal fenestra (Figure 10g-h); internal septa occurring at the base of every branch..................11
10. Dorsal annulus is not so extensive; septa rare or absent..................12

11. Floatoblast dorsal valve nearly flat, suture between valves visible in dorsal view (Figure 21c); sessoblast surface uniformly granular; tentacles about 39...................... *Plumatella emarginata* (p. 37).
11. Floatoblast valves almost equally convex; sessoblast surface with network of dark lines (Figure 23b); tentacles about 32.....................................
.... *Plumatella reticulata* (p. 40).

12. Floatoblast dorsal fenestra length more than 1.5 times width; colony compact (Figure 8e-f); zooids attached to substrate or to each other along most of their length; tentacles fewer than 30.....................................
.... *Plumatella casmiana* (p. 28)
12. Floatoblast dorsal fenestra length less than 1.5 times width; colony form variable; tentacles more than 30..................13

13. Floatoblasts asymmetrical in side view (Figure 19d); zooids sometimes attached to each other throughout their length to form a solid cushiony mass (Figure 19a); septa common.....................................
.... *Plumatella fungosa* (p.34)
13. Floatoblast valves almost equally convex and symmetrical in side view (Figure 17d); colony walls usually transparent; colony never forming a solid mass; septa absent.....................................
.... *Plumatella repens* (p. 31).

14. Polypides in groups of 2; cystid often wrinkled; zooids bent sharply in various directions (few reports only from Michigan and Pennsylvania)...*Stollela indica.*
14. Polypides in groups of 2 to 5; clusters oriented vertically to substrate, zooids not bent sharply (known in North America only from Michigan)...*Stollela evelinae.*

15. Colony becoming large, with more than 50 zooids per square centimeter (Figure 8h); mouth region with red pigmentation; prominent pair of white spots on lophophore; statoblasts with hooked spines radiating from annulus (Figure 10a). PECTINATELLIDAE..
.... *Pectinatella magnifica* (p. 49).
15. Colony smaller, mouth region not pigmented, statoblasts otherwise.......16

16. Statoblast with hooked spines radiating beyond periphery from capsule (Figure 10b), colony linear. CRISTATELLIDAE..
.... *Cristatella mucedo* (p. 52).
16. Statoblast with row of small hooks at each end (Figure 10c); colonies more globose than linear. LOPHOPODIDAE..
.... *Lophopodella carteri* (p. 46).

DESCRIPTION OF OHIO BRYOZOANS

CLASS PHYLACTOLAEMATA

Family Fredericellidae

Fredericella indica Annandale, 1909

DESCRIPTION OF OHIO SPECIMENS. The colony is a seemingly disorganized tangle of slender tubules, dichotomously branched, some attached to the substrate and others largely free and erect (Figure 11a). Incrustation varies from a light dusting to a thick covering of particles. The color is pale yellow to brown. Adherent branches may bear a very distinct keel and furrow, but free branches are mostly terete and lack both keel and furrow. The tentacular crown is circular in outline rather than crescentic, and the tentacles are somewhat shorter than in other phylactolaemate bryozoa (Figure 11b-c).

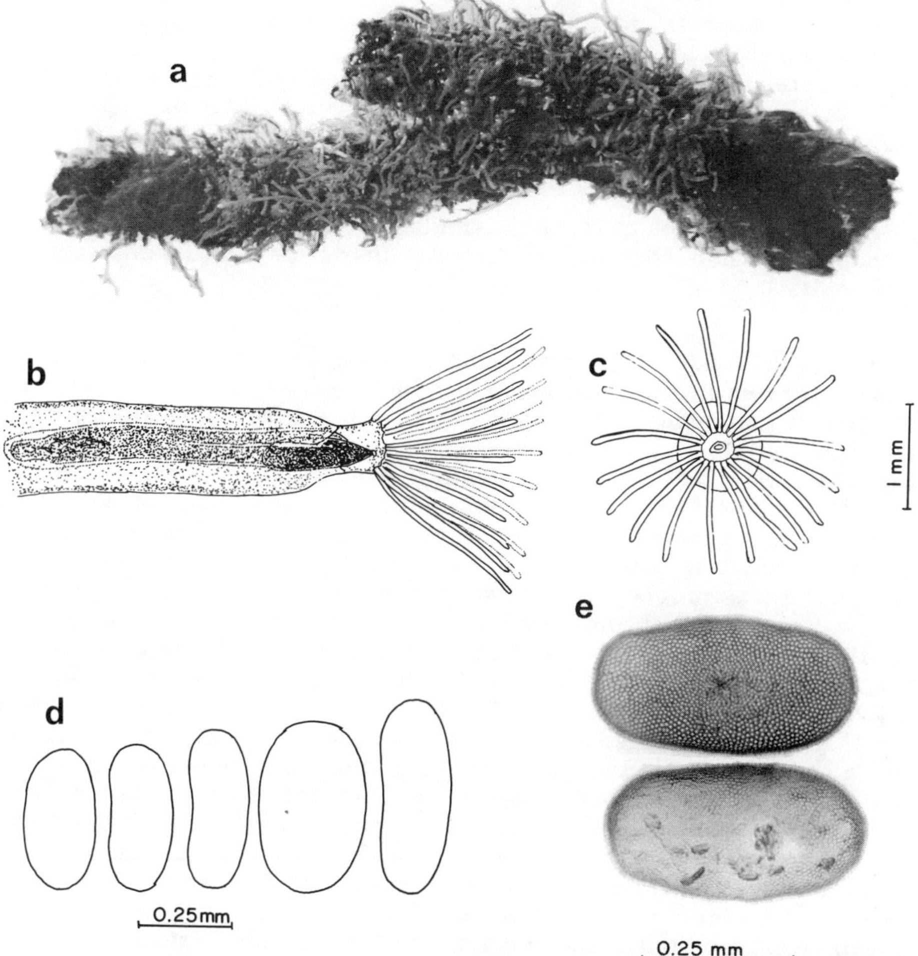

Figure 11. *Fredericella indica*. a—Colony growing on a twig; b—single zooid; c—frontal view of lophophore; d—range of variation in statoblast dimensions; e—separated valves of the statoblast, showing extensive pitting on both valves, with the ventral valve below, dorsal on top.

21

The statoblast is a piptoblast only, bearing no annulus and having no buoyancy. While not cemented directly to the substrate, it often does adhere to the substrate side of the colony wall by means of an oval formation of small, keel-like processes extending from the ventral valve (Figure 3g). The statoblast outline varies from nearly round to elongate and somewhat rectangular (Figure 10d). An asymmetrical kidney shape is not uncommon. Toriumi (1951) found that statoblast dimensions correlate with the size of the space in which it is found which, in turn, may be influenced by environmental conditions.

All mature statoblasts are minutely pitted on the surfaces of both valves. This is the result of a fine reticulum of continuous ridges forming open, polygonal cells (Figures 3f-g; 11e). This pattern produces a dull, velvety or granular appearance when the statoblast is dry and viewed with reflected light, but this feature is difficult to see when the statoblast is wet.

Similar species:

Fredericella sultana—statoblasts appear smooth even when dry; unknown in North America.

Fredericella crenulata—statoblasts are rough-textured, with tiny denticles along the margins; described only from a lake in Brazil (Marcus, 1946).

Fredericella australiensis—zooids somewhat larger than in *F. indica*; statoblasts are smooth, more rounded, and generally much more abundant (see description on p. 24).

Plumatella fruticosa—colony form is similar to *Fredericella indica,* but lophophore is horseshoe-shaped and statoblasts are true floatoblasts and sessoblasts.

TAXONOMY. In 1779, Blumenbach published the name *Tubularia sultana* for a bryozoan collected in Germany, which he believed to be related to marine hydrozoans of that genus. The lophophore bore tentacles arranged in a circle. Later, after the bryozoans had become recognized as distinct from the hydrozoans, Gervais (1838) changed the genus to *Fredericella,* honoring the French naturalist, Frederick Cuvier. Originally considered a member of the genus *Plumatella,* the separate family, Fredericellidae, now is recognized to include all phylactolaemate bryozoans having piptoblasts and a circular lophophore.

Many workers consider the common *Fredericella* species in North America to be *F. sultana.* However, piptoblast valves of the latter are perfectly smooth, as illustrated by Allman (1856) and confirmed by the SEM photos of Mundy (1980) and Geimer and Massard (1986). These contrast sharply with the extensively and deeply pitted piptoblasts valves found on North American species (Figure 10e). Such a rough surface texture was first noted by Annandale (1909) in specimens collected from western India, which he named *F. indica.* Rogick saw the pitted statoblasts valves in Lake Erie colonies, but believed them to be the result of age (1937a). A re-examination of her material has shown this not to be the case (Wood, 1979). In Bushnell's extensive survey of Michigan material (1965a), the textured surface of *Fredericella* piptoblasts was always "obscure to reasonably clear," implying the existence of forms intermediate between rough and smooth types. Similar observations have been made by Rao (1973) in India and by Bonneto and Cordiviola de Yuan (1965) in South America. However, I have yet to find a piptoblast of this species in North America which, when properly examined, was not heavily and uniformly pitted. Perhaps other workers have made their examinations only when the piptoblasts were wet, when a film of moisture masks the rough texture.

As in the cases of other bryozoan species, many features of *Fredericella* colonies are influenced by environmental conditions. According to Toriumi (1951) such features include shape and size of the zooecium, shape and size of statoblasts, and colony growth form (dense or sparse, zooids erect or recumbant). Many of the dozen or so *Fredericella* species proposed in the past were based on these environmentally variable features, and are recognized no longer (Table 2).

Table 2. *Fredericella indica.* Measurements (dimensions in micrometers, standard deviations shown in parentheses).

ITEM	MAXIMUM	MINIMUM	MEAN	NUMBER OF MEASUREMENTS
Number of tentacles	23	16	19.4 (1.4)	82
Zooecial diameter	382	192	—	32
Statoblast length	515	351	434 (22.1)	70
Statoblast width	281	163	217 (20.6)	70
Statoblast length/ width	2.9	1.3	2.03 (0.22)	70

BIOLOGY. *Fredericella indica* is distributed widely throughout Ohio (Figure 12). Although not uncommon, it forms small, diffuse colonies which are overlooked easily. Substrates usually are large rocks and heavy, old logs which maintain their position from year to year. However, some Ohio colonies also occur on leaves of lotus (*Nelumbo lutea*). Bushnell (1966) reports a few occurrences in Michigan on the yellow pond lily, *Nuphar advena.*

Fredericella indica tolerates a very wide range of temperature. Bushnell (1966) found living colonies on substrates beneath the ice of Gull Lake from December through March in water of 1.5° to 2° C.

The relatively small number of statoblasts produced by the colonies never are released freely, and the statoblasts alone probably do not serve as disseminules. The effective unit of dispersal is probably either the motile phase (larva) or a fragmented piece of the colony (Wood, 1973).

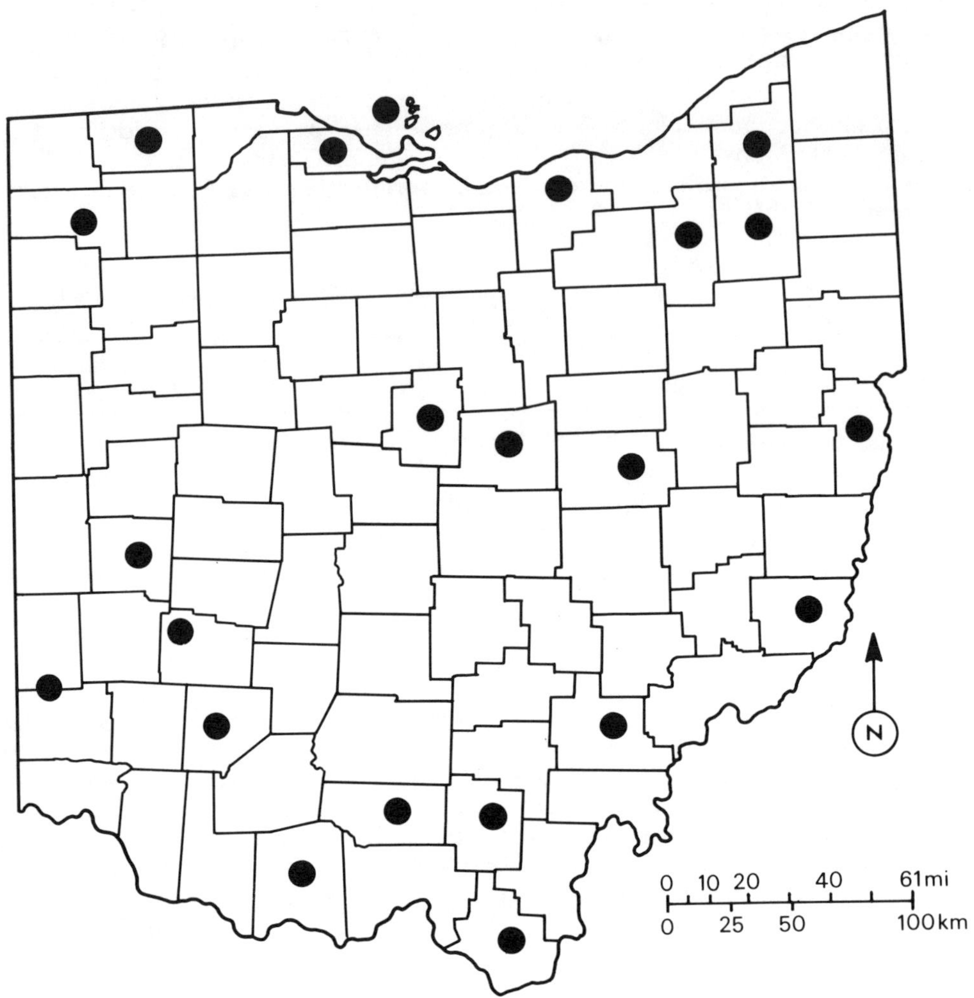

Figure 12. Ohio distribution of *Fredericella indica*.

Fredericella australiensis Goddard, 1909

DESCRIPTION OF OHIO SPECIMENS. Colonies are composed of branching tubules, somewhat more robust than those of *F. sultana*. They can be compact to very diffuse, with branches ranging from entirely adherent to largely free of the substrate (Figure 13a). Adherent portions show a distinct keel and furrow, while free branches are generally terete and without these features. The colony wall is encrusted and generally opaque.

Statoblasts are piptoblasts only, neither cemented to the substrate nor equipped with a buoyant ring. They are produced continuously and in great number, often lying end-to-end within the branches. Their shape is round to oval, seldom elongate (Figure 13b-c). Individual valves clearly exhibit a uniformly thickened rim when observed under the microscope.

Similar species:

Fredericella indica–similar in colony morphology; has smaller zooecial diameter; statoblasts are much less abundant and more elongate; and statoblast valves appear densely pitted when examined dry.

Fredericella sultana–statoblasts much less abundant and more elongate; unknown so far in North America.

Plumatella fruticosa–long branches free and often tangled; statoblasts are true floatoblasts and sessoblasts; lophophore horseshoe shaped.

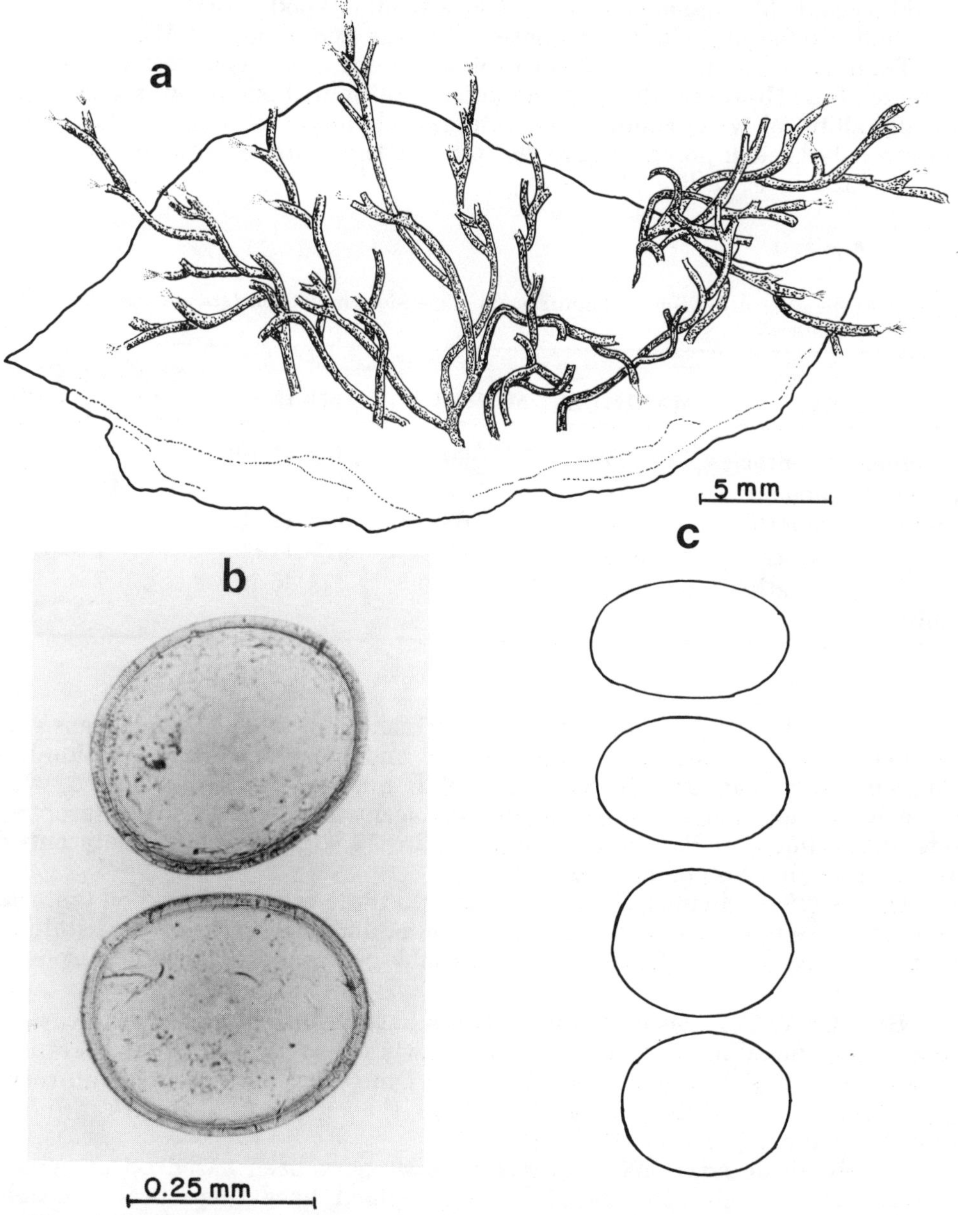

Figure 13. *Fredericella australiensis.* a–Sparse colony growing on a stone; b–separated valves of the statoblast, showing thick rims and smooth surfaces; c–range of variation in statoblast dimensions.

TAXONOMY. *Fredericella australiensis* was first reported and described in 1909 by Goddard from water supply reservoirs in New South Wales, Australia. Subsequent discoveries of this species were made from the following disparate locations:

Tunisia: Oued Abid (Borg, 1936)
Wyoming: Uinta County (Rogick, 1945);
Bolivia: Lake Titicaca (Marcus, 1953);
Belgium: La Voer (Lacourt, 1968);
Mexico: Chihuahua and Durango States (Bushnell, 1971);
Maryland: Montgomery County (Backus and Wood, 1981);
Ohio: Preble and Clinton Counties (Backus and Wood, 1981).

There is not perfect agreement in descriptions of the specimens from any of these sites. However, the key characters are unmistakable: piptoblasts are broadly elliptical, wide-rimmed, smooth, and abundant; zooecia are large and not strongly keeled; polypides are shorter and the tentacle number somewhat greater than in *F. indica* (Table 3).

Table 3. *Fredericella australiensis.* Measurements (dimensions in micrometers, standard deviations shown in parentheses).

ITEM	MAXIMUM	MINIMUM	MEAN	NUMBER OF MEASUREMENTS
Number of tentacles	25	20	21.6 (1.42)	22
Zooecial diameter	481	289	—	40
Statoblast length	429	368	369 (17.2)	7
Statoblast width	313	253	277 (19.2)	7
Statoblast length/ width	1.7	1.2	1.3 (0.16)	7

A study by Backus and Wood (1981) showed that *F. australiensis* is distinct from *F. indica* both karyotypically and morphologically (in side-by-side laboratory rearing). The karyotpye of *F. australiensis* shows 2n=16, with three large metacentric, three smaller metacentric, and two submetacentric sets. In specimens of *F. indica,* by contrast, 2n=14 with three large metacentric and four smaller submetacentric sets.

Rogick (1945) divided *F. australiensis* into three varieties based on tentacle number, cross-sectional shape of the zooecium, and the degree of incrustation. However, each of these features is too variable for useful diagnostic purposes.

BIOLOGY. Colonies of *F. australiensis* have been found in a wide variety of habitats, including both standing and slowly flowing water, in temperatures ranging from near-freezing to 30°-35° C. In Ohio, they have occurred on rocks and submerged wood (Figure 14). Elsewhere, the substrates also have included rooted aquatic vegetation (Bushnell, 1968).

Nearly all descriptions of *F. australiensis* have been based on preserved material from single samples. Only in Maryland have there been multiple collections over a period of years (Backus, personal communication). Field observations, so far, suggest that the colonies undergo periods of rapid growth followed by sudden senescence and a relatively long statoblast dormancy. Although it can be found with other bryozoans, it appears to flourish most when colonies of other species are either small or absent. Living material in

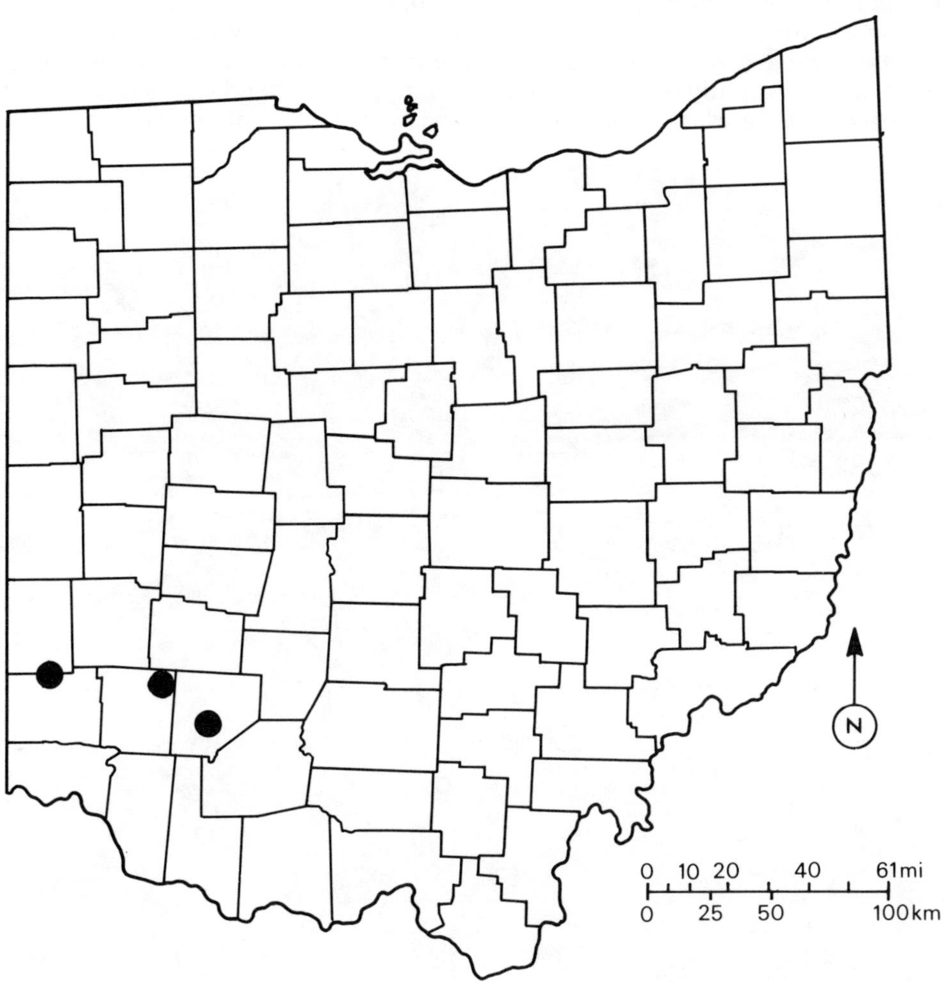

Figure 14. Ohio distribution of *Fredericella australiensis.*

Ohio has been found only during the summer and at only two sites (Figure 14). In Maryland it has been collected also during late autumn, winter, and early spring.

Family Plumatellidae

Plumatella casmiana Oka, 1907

DESCRIPTION OF OHIO SPECIMENS. The colony is compact with short, regularly branching tubules entirely adherent either to the substrate or to each other. There are no free branches. Zooids generally are raised slightly only at the tips. Where the substrate is limited, as on a small twig or stem, densely crowded zooids may become fully erect and firmly cemented together as if possessing common walls. This condition also may occur in the middle region of any rapidly growing colony. On the other hand, on relatively unrestricted substrates, the colonies form dense, mat-like patches with branches

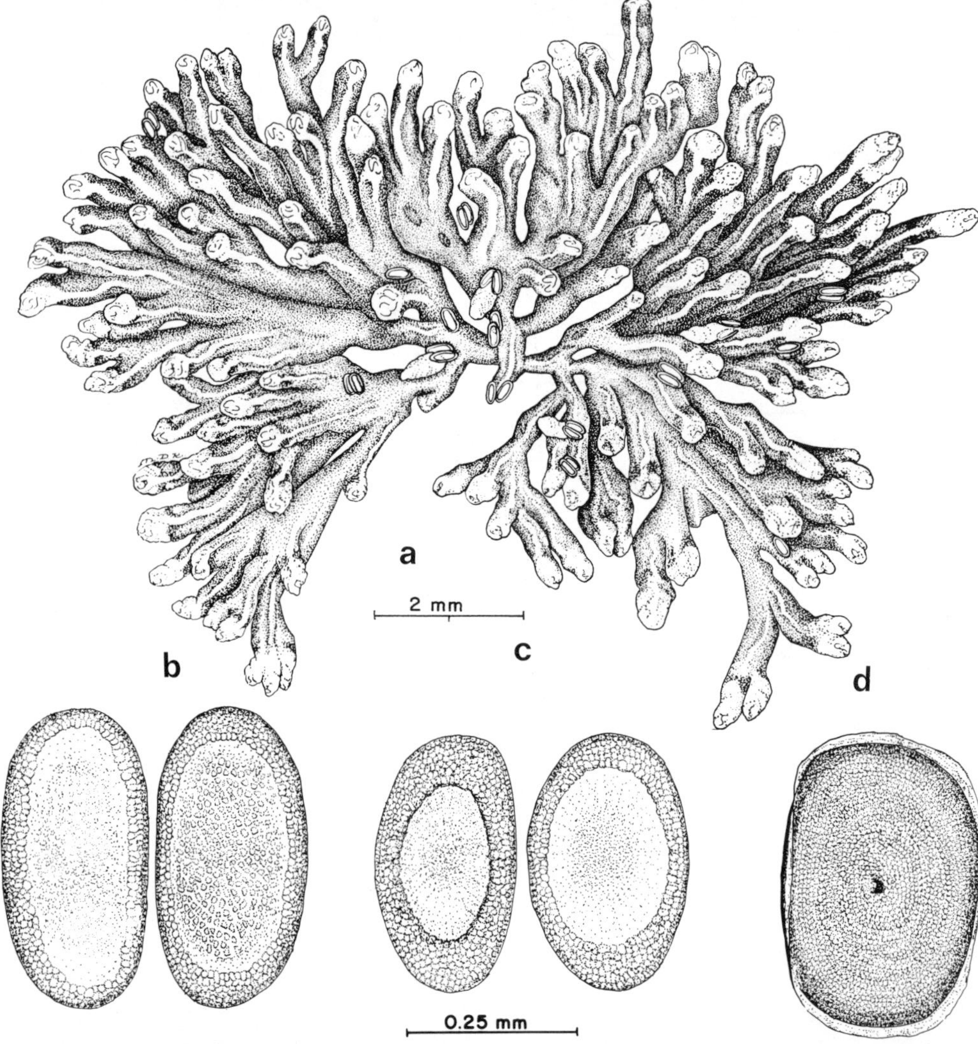

Figure 15. *Plumatella casmiana*. a—Colony with several released leptoblasts (from Rogick, 1941); b-d—typical statoblasts valves of *Plumatella casmiana* in Ohio: b—leptoblast, c—capsuled floatoblast, d—sessoblast, showing narrow lamella.

28

radiating from common points of origin. Branches may grow briefly in parallel series, but rarely do they overlap. Zooids bear a distinct keel, furrow, and emargination. The ectocyst is rather brittle and usually is well-incrusted.

Statoblasts occur in three distinct classes:

leptoblasts (Figure 15b). Delicate, elongate, thin-walled statoblasts with a uniformly narrow annulus. The dorsal and ventral valves are very similar and separate easily upon release from the colony. The transparent central area of each valve is sparsely tuberculated;

capsuled floatoblasts (Figure 15c). Floatoblasts of the normal plumatellid type with a pneumatic annulus encroaching upon the smooth oval capsule more on the dorsal than on the ventral valve, typical of most plumatellid floatoblasts, and;

sessoblasts (Figure 15d). Broadly oval and cemented firmly through the colony wall to the substrate. The annulus is much more narrow than in either *Plumatella repens* or *P. emarginata.* The frontal capsule (facing away from the substrate) bears a central, raised tubercle encircled by fine concentric lines. The overlying periblast sometimes also has a similar tubercle.

Although several other types of statoblasts have been described for this species (Wiebach, 1968; Rao et al, 1978) none has been found in Ohio specimens, other than those listed above.

Plumatella casmiana is one of the most distinctive species among the Plumatellidae. Key features include:

1. Presence of leptoblasts—these are found in no other species of bryozoan;
2. A relatively small number of tentacles;
3. Sessoblast with only a very narrow annulus; frontal capsule with a single, raised central tubercle; frontal valve mostly smooth, not granular;
4. In living colonies, the rhythmic flicking of lophophore tentacles (see below under "Biology"), and;
5. Colony morphology: dense and usually very flat, mat-like; becoming spindle-shaped around thin twigs or stems.

TAXONOMY. *Plumatella casmiana* was first described in 1907 by Oka from Kasumiga-Ura, a lake near Tokyo. The species' major distinguishing feature was the thin-walled leptoblast. In fact, it was solely by means of this atypical statoblast that the species was identified later in other collections.

Specimens of *P. casmiana* lacking leptoblasts were known for a time by other names, including *P. repens* var. *flabellum* (Rogick, 1935a) and *P. annulata* (Hôzawa and Toriumi, 1940). The discovery by Toriumi (1942b) of both leptoblasts and capsuled floatoblasts in the same colony led to a unification of these names under *Plumatella casmiana.* Since then, additional statoblast types have been attributed to *P. casmiana,* although apparently none is as common or as distinctive as the leptoblast and capsuled floatoblast.

BIOLOGY. *Plumatella casmiana* is nearly cosmopolitan in its distribution, now reported from every continent except Australia. It is found in small, pond-like habitats throughout Ohio (Figure 16). In the area of western Lake Erie, for example, *P. casmiana* seldomly is found in the lake itself, but mainly in small island ponds. Common substrates are wood, stone, aquatic vegetation, old rubber tires, and glass bottles, with no apparent specificity.

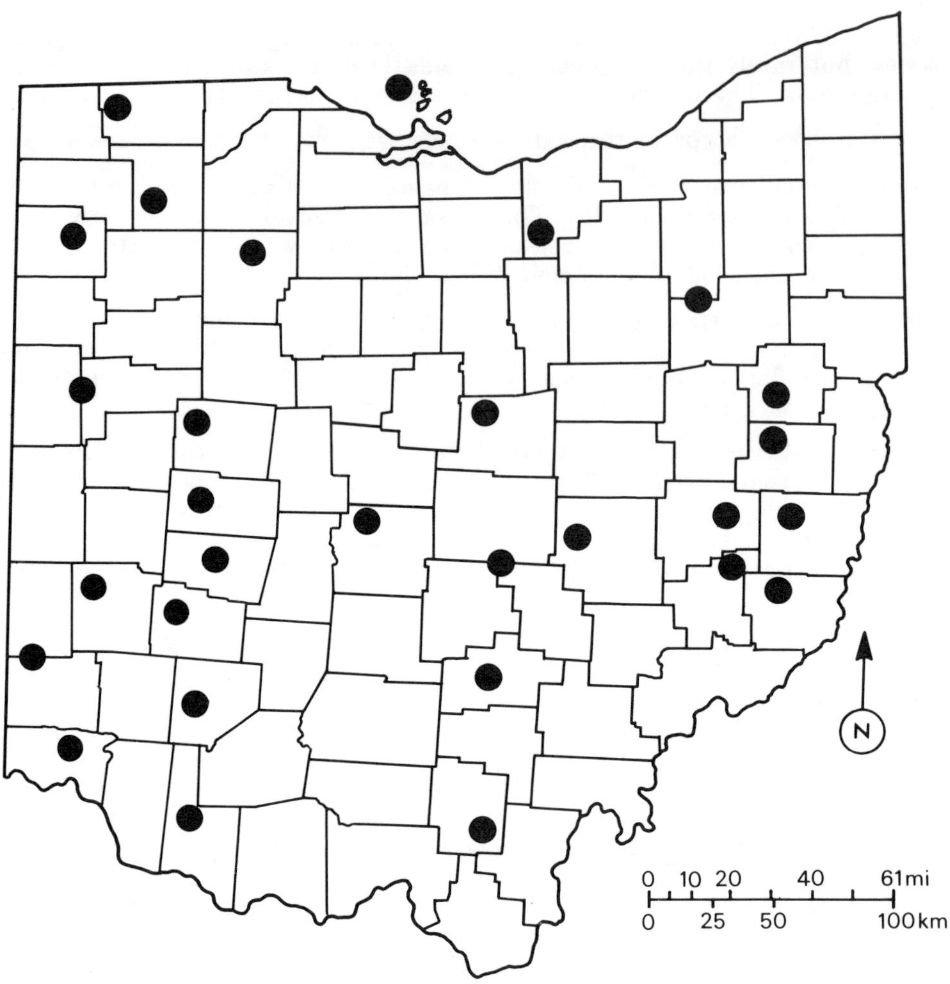

Figure 16. Ohio distribution of *Plumatella casmiana*.

Probably the most biologically unique feature of *P. casmiana* is the formation of several types of floatoblasts, notably the thin-walled leptoblast. As a rule, the phylactolaemate statoblast contains a small mass of germinal tissue and large stores of yolky food reserves. In this case, however, the statoblast is released with few yolk reserves and a fully formed polypide. It germinates immediately, with the ancestrula often attaching to the parent colony (Figure 15a). Survival rates for leptoblast ancestrulae are relatively low but the mechanism nonetheless enables this species to achieve rapid population growth during the summer months.

In addition to leptoblasts, *Plumatella casmiana* in Ohio produces the more typical plumatellid capsuled floatoblasts (Table 4). From her work with specimens from Lake Erie, Rogick (1943) found leptoblasts most abundant in the early summer, with capsuled floatoblasts appearing in July and August. She noted that both floatoblast types sometimes were present in the same colony. These observations essentially were corroborated in Italy by Viganò (1968), in Colorado by Wood (1973), and in Japan by Toriumi (1955a) and Mukai

Table 4. *Plumatella casmiana.* Measurements (dimensions in microns, standard deviations shown in parentheses; floatoblast data for capsuled type only).

ITEM	MAXIMUM	MINIMUM	MEAN	NUMBER OF MEASUREMENTS
Number of tentacles	45	26	34.6 (3.5)	71
Floatoblast length	505	306	379 (48.3)	101
Floatoblast width	281	163	225 (22.1)	101
Floatoblast length/ width	2.8	1.3	1.7 (0.19)	101

et al (1983), with allowance for differences in climate. Specimens collected throughout Ohio show the first capsuled floatoblasts appearing in July, while leptoblasts occur from May through September. Colonies reared in the laboratory usually form leptoblasts only, regardless of the time of year.

A second feature peculiar to *Plumatella casmiana* is a rhythmic, synchronous flicking of the lophophore tentacles. First noticed by Viganò (1968), this behavior since has been observed by others in both wild and laboratory-reared colonies. The beat begins shortly after emergence of the new zooid and continues for several days. It may serve to improve the rate of gas exchange at the tentacles. This behavior, which is so typical of *P. casmiana,* has been seen elsewhere only in *Plumatella reticulata,* and for all practical purposes, it may be considered a reliable taxonomic character in living specimens.

Sessoblasts in this species are generally uncommon during most of the season, but are formed in large numbers in the late autumn. Before the onset of winter, disintegrating colonies of *P. casmiana* reveal branching chains of sessoblasts cemented end-to-end on the substrate.

Larvae generally are released in Ohio during late June and July. Each carries two zooids.

Plumatella repens (Linnaeus, 1758)

DESCRIPTION OF OHIO SPECIMENS. The colony consists of tubular branches which, on an unrestricted substrate normally are long and linear, hence the name *repens,* from the Latin *reptans* = spreading (Figure 17a). Occasionally, dense colony growths are encountered in which elongated zooids are aligned vertically suggesting the thick pile of a carpet. A relatively consistent feature of the colony wall is the absence of heavy incrustation. The cuticle varies in color from pale yellow - almost clear - to a dark reddish brown, but seldom is it rendered opaque from the tiny incrusted particles so common among other plumatellids. Compared to those of other species, the tentacles are spectacularly long and delicate-looking, although these are so difficult to measure accurately that they are a poor diagnostic feature.

Floatoblasts are usually numerous and easily visible within the colony. They are broadly oval, with considerable variation in the size of the dorsal fenestra (Figure 17b-c, Table 5). In side view both valves appear about equally convex (Figure 17d). Under high magnification, it can be seen that each cell in the fenestra of the isolated dorsal periblast contains a small tubercle, giving this portion of the statoblast a granular appearance when dry (Figure 4). The sessoblast has a highly tuberculated surface and well-developed annulus, or lamella, with a minutely serrated outer edge (Figure 17e). The width of the lamella and its almost lateral insertion of the statoblast are diagnostic features of this species. However, when the sessoblast is dry, its lamella assumes a more oblique angle similar to the sessoblast of *P. emarginata*.

Similar species:
Plumatella fungosa—floatoblasts are distinctly asymmetrical in side view with
flat dorsal valve and strongly convex ventral valve; dark septa common;
zooids in older colonies form solid mass.

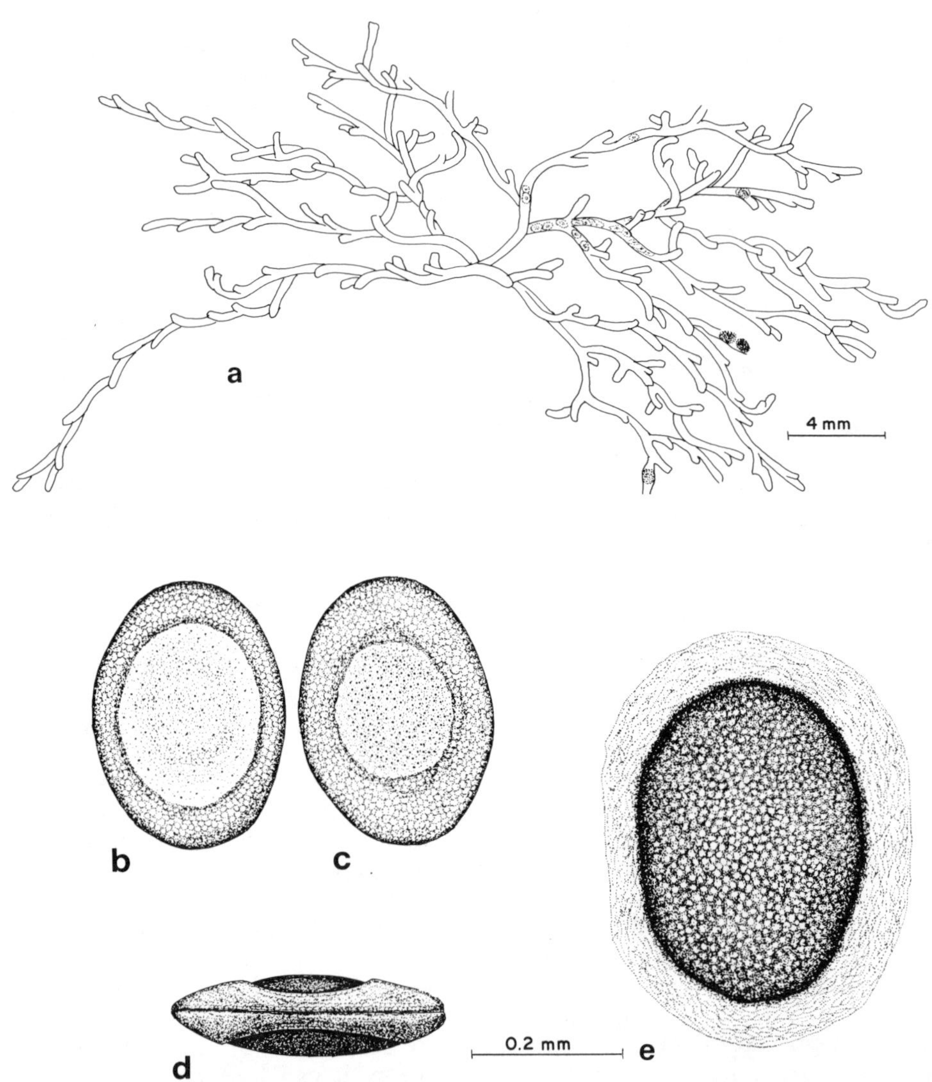

Figure 17. *Plumatella repens*. a-typical colony form; b-ventral valve of floatoblast; c-dorsal valve of floatoblast; d-lateral view of floatoblast showing equally convex valves; e-sessoblast, showing typically wide lamella.

TAXONOMY. Originally called *Tubipora repens* by Linnaeus, Allman renamed the species *Plumatella repens* in 1844 as part of his new classification scheme. For a while, all *Plumatella* species were considered variations of *P. repens,* and the name was applied to any branching colony with horseshoe-shaped lophophores. Recently, as these variations have been elevated to species in their own right, debate has centered on the degree of variability within the genus.

BIOLOGY. *Plumatella repens* is probably the most widely distributed phylactolaemate species, occurring on every continent. It grows best in quiet or slowly moving water, and frequently is found in small ponds. In larger bodies of water, poor colony growth has been linked to vigorous wave action (Jónasson, 1963). In Ohio, colonies have been collected mostly from submerged wood subtrates, although glass, rock, and several macrophyte species have supported colonies as well (Figure 18).

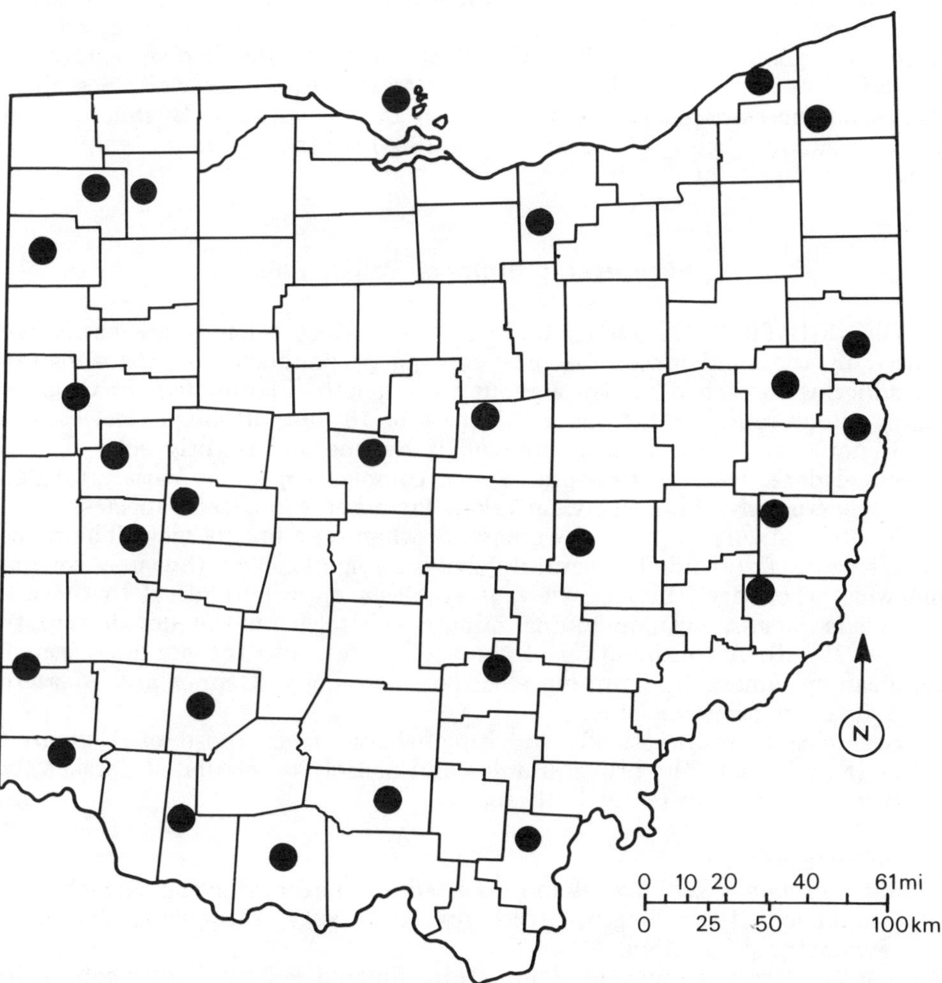

Figure 18. Ohio distribution of *Plumatella repens.*

33

Table 5. *Plumatella repens*. Measurements (dimensions in microns, standard deviations shown in parentheses).

ITEM	MAXIMUM	MINIMUM	MEAN	NUMBER OF MEASUREMENTS
Number of tentacles	59	28	42.0 (7.9)	78
Floatoblast length	418	311	366 (18.2)	400
Floatoblast width	321	209	254 (16.3)	400
Floatoblast length/ width	1.76	1.08	1.45 (0.10)	400

In Ohio, statoblasts of *P. repens* are among the first to germinate in the spring, when water temperatures reach 8°-10°C. Further warming of the water encourages rapid colony growth. Gametogenesis begins in early May, with larval release occurring for no more than a few weeks in June. Statoblast production ensues, with as many as eight floatoblasts formed by a single zooid. In Michigan, Brown (1933) counted 57 floatoblasts per cm^2 of occupied substrate. As a colony becomes filled with floatoblasts, its zooids disintegrate and the statoblasts are released through fissures in the colony wall. Sessoblast production increases as temperatures drop in the autumn. Bushnell (1966) found no budding below 9°C, and Brown (1933) set the minimum temperature of living *P. repens* at 6°C.

Plumatella fungosa Pallas, 1768

DESCRIPTION OF OHIO SPECIMENS. Most colonies are easily recognized by the solid mass of zooids growing perpendicular to the substrate and adhering to each other throughout their length (Figure 19a). In Ohio the largest colony found so far has a thickness of 18 mm, although elsewhere 50 mm is not uncommon. The colony wall is hyaline and slightly colored, with occasional dark, incomplete septa. Young colonies appear as loosely tangled, slightly encrusted zooids easily mistaken for other *Plumatella* species.

Floatoblasts are strongly asymmetrical when seen in side view. The dorsal valve is nearly flat, while the ventral valve is strongly convex (but may collapse somewhat when dry). In face view, floatoblasts appear identical to those of *Plumatella repens*, even including minute tubercles on the dorsal fenestra (Figures 19b-d). A continuation of these tubercles onto the annulus, seen in European specimens by scanning electron microscopy (Geimer and Massard, 1987) does not occur in Ohio.

Sessoblasts are practically indistinguishable from those of *Plumatella repens* (Figure 19e). The broad annulus and densely tuberculated dorsal valve are characteristic features in both species.

Similar species:

Plumatella repens—zooids seldom encrusted and never adhering to each other throughout their length; septa absent or very infrequent; floatoblast symmetrical in lateral view.

Plumatella casmiana—may assume a thin, fungoid colony form when zooids are crowded, but zooids have a distinct keel, floatoblasts are easily recognized, and tentacles are generally fewer than in *P. fungosa*.

TAXONOMY. *Plumatella fungosa* is generally recognized on the basis of its striking fungoid colonies with zooids completely adherent throughout their length. However, in two detailed studies, Toriumi (1971a; 1971b) could find no significant differences between *P. fungosa* and *P. repens*. His comparison of laboratory-reared colonies was especially persuasive, seeming to confirm the widely held belief that *P. fungosa* is merely a robust form of *P. repens*. However, a study by Mundy and Thorpe (1979) using electrophoresis suggested genetic differences between single specimens of the two species. These authors, as well as Giemer and Massard (1987), have also described minute details of

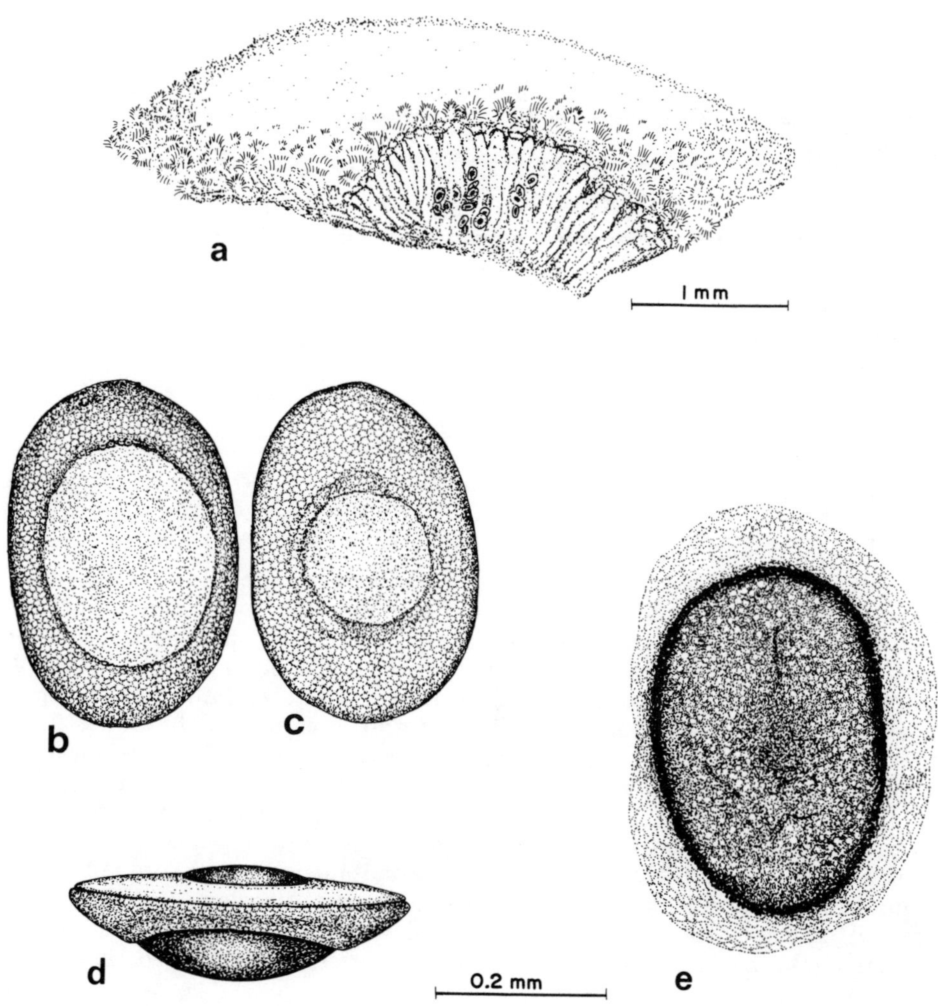

Figure 19. *Plumatella fungosa*. a-small fungoid colony with one edge broken off to reveal parallel series of vertical zooids, all firmly adherent to each other; b-ventral valve of floatoblast; c-dorsal valve of floatoblast; d-lateral view of floatoblast showing typical asymmetry; e-sessoblast.

the statoblast surface which are thought to distinguish the species. Those details are best detected with scanning electron microscopy, and even then they are not reliable for North American species.

Possibly a more significant taxonomic feature in *Plumatella fungosa* is the strong asymmetry of the floatoblast in lateral view (Figure 19d). The dorsal orientation of the suture due to a flat dorsal valve is clearly illustrated by Allman (1856), Dumortier and Van Beneden (1850), and Brien (1953). By contrast, floatoblasts of *P. repens* are almost equally convex and symmetrical in lateral view. While floatoblast asymmetry is known in other species (e.g., *Plumatella emarginata*, *Stollela indica*, and *Hyalinella carvalhoi*), these species are easily distinguished by other features.

Further work is needed to determine the range of variation in the morphology of *P. fungosa* statoblasts. For now it may be considered a valid species distinct from *P. repens*.

Table 6. *Plumatella fungosa*. Measurements (dimensions in microns, standard deviations shown in parentheses).

ITEM	MAXIMUM	MINIMUM	MEAN	NUMBER OF MEASUREMENTS
Number of tentacles	40	38	39 (0.74)	11
Floatoblast length	440	350	409 (16.9)	30
Floatoblast width	310	230	290 (17.1)	30
Floatoblast length/ width	1.8	1.3	1.4 (0.09)	30

BIOLOGY. *Plumatella fungosa* is widely distributed in North America, Europe, and Northern Asia. In Europe it is among the most common bryozoan species, while in North America it is reported from fewer than a dozen scattered sites in Maine, Massachusetts, Ohio, Illinois, and Colorado (Figure 20). Eutrophic ponds and protected river backwaters are favored habitats. The species tolerates a high level of organic pollution and is estimated by Job (1976) to remove significant amounts of nitrogen from the water. Substrates tend to be solid structures of wood, rock, masonry, or plastic, rather than aquatic plants.

While most young *Plumatella* zooids are somewhat sticky, those of *P. fungosa* apparently retain this quality longer than in other species. As a result, zooids crowded by limited substrate or by simultaneous germination of adjacent statoblasts become cemented firmly along their entire length. Where zooids are not crowded, they are often encrusted with particles from the water. In this case, the zooids may eventually become richly intertwined, but not cemented together.

Two workers have observed microsporidian parasites in *P. fungosa* (Schröder, 1913; Braem, 1911). This is probably not uncommon among other *Plumatella* species as well. A close association between *P. fungosa* and the cyanophyte, *Phormidium fragile*, was noted by Trauberg (1940), possibly similar to the relationship between *Hyalinella punctata* and *Oscillatoria* spp.

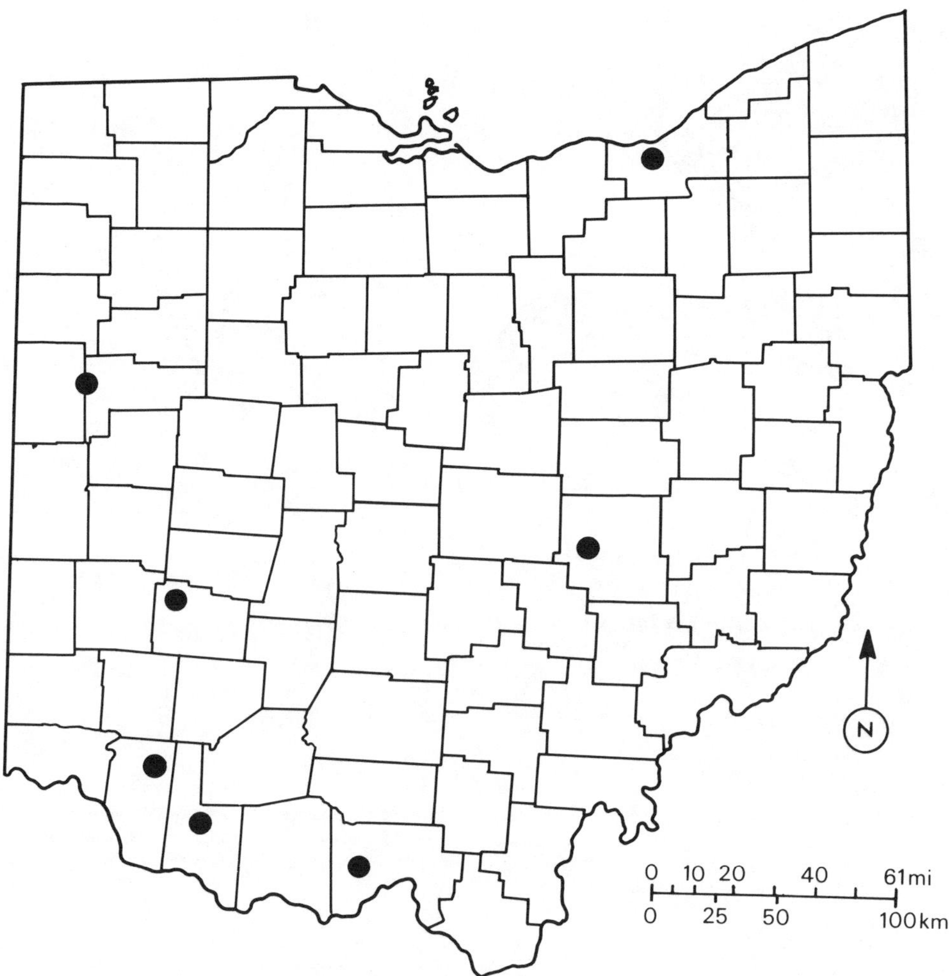

Figure 20. Ohio distribution of *Plumatella fungosa*.

Plumatella emarginata Allman, 1884

DESCRIPTION OF OHIO SPECIMENS. Morphologically, this is the most variable of Ohio's freshwater bryozoans. The specific name *emarginata* refers to the tips of the zooids where the colony wall is hyaline, as though it were protruding from an opaque, encrusted sheath. Unfortunately, not all specimens show this feature, and those that do share it with virtually every other species of *Plumatella* with an encrusted ectocyst.

Colonies tend to be rather dense and somewhat ragged-looking, with zooids bending away from the substrate. Those colonies which are crowded or on loose substrates may send out free branches, not unlike those of *Fredericella* spp. or *Plumatella fruticosa*. Shrivastava and Rao (1985) have found, in India, zooids compressed in a honeycomb-like form of growth, but this has not been observed in Ohio specimens. Toriumi (1952b) showed from laboratory rearing that colony form is directly related to budding rate. Zooecia are nearly always

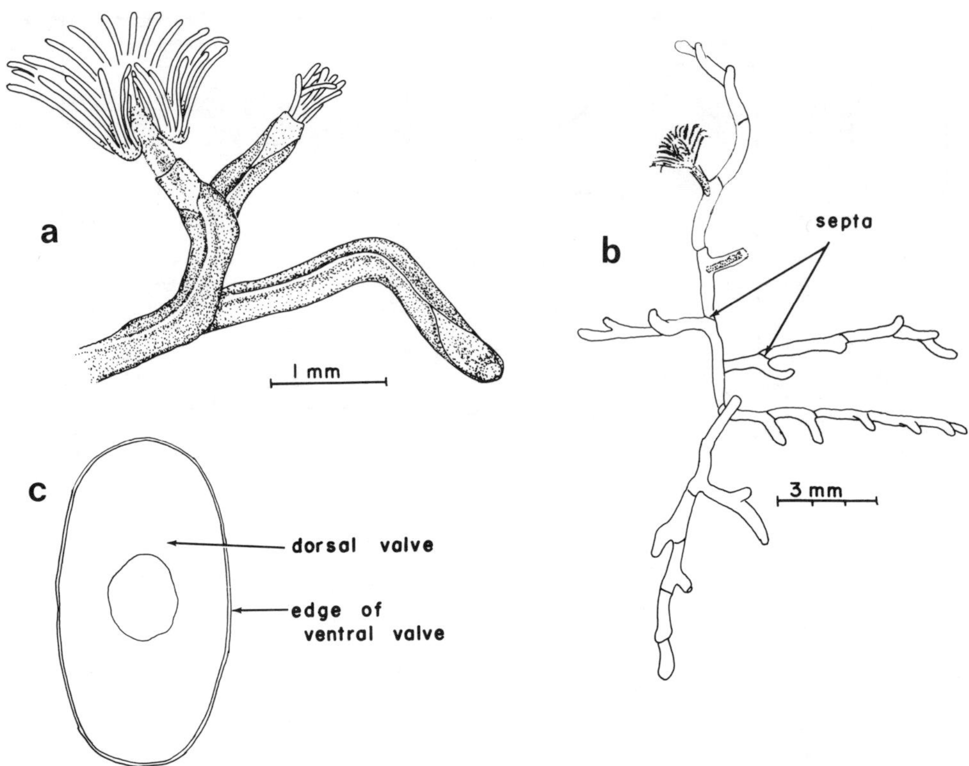

Figure 21. *Plumatella emarginata*. a—Three zooids, showing strong keel and emargination; b—young laboratory-reared colony, showing position of septa between zooids; c—dorsal view of floatoblast showing ventral valve extending beyond the dorsal margin (see Figure 3a-e for more detailed views of statoblasts).

well encrusted, exhibiting both keel and furrow on horizontal branches (Figure 21a). Septa are common, occurring at the junction of every branch (Figure 21b).

Floatoblasts are distinctive. Although dimensions vary considerably, the general shape is elliptical (Figures 3a-c). The dorsal valve is flat and bears an extensive annulus that leaves only a small fenestra. The ventral valve is strongly concave, with an outer rim extending beyond the edges of the dorsal valve, such that the suture is clearly visible from the dorsal view (Figure 21c).

In the sessoblast, the dorsal periblast has a lamella which is variable in width, sometimes lightly reticulated, with margins either smooth or minutely serrated. The dorsal valve is covered with minute tubercles.

TAXONOMY. Since *Plumatella emarginata* was described by Allman in 1844, its status has not been challenged seriously. However, in view of its similarity to the recently described *Plumatella reticulata*, it is possible that these two have been confused by previous workers and treated as a single species. Bushnell (1965b) describes a form from Michigan in which consistent features from several sites appear intermediate between *P. emarginata* and *P. repens*. There may well be some hybridization between these and other species of *Plumatella*.

Table 7. *Plumatella emarginata.* Measurements (dimensions in micrometers, standard deviations shown in parentheses).

ITEM	MAXIMUM	MINIMUM	MEAN	NUMBER OF MEASUREMENTS
Number of tentacles	42	21	33 (4.3)	107
Floatoblast length	515	265	424 (42.7)	158
Floatoblast width	281	204	239 (16.2)	158
Floatoblast length/ width	1.9	1.3	1.8 (0.07)	158

ECOLOGY. *Plumatella emarginata* is distributed widely, occurring on every continent. It is considered a species of rivers and streams because, although it thrives in lentic environments, it also tolerates the turbulence of

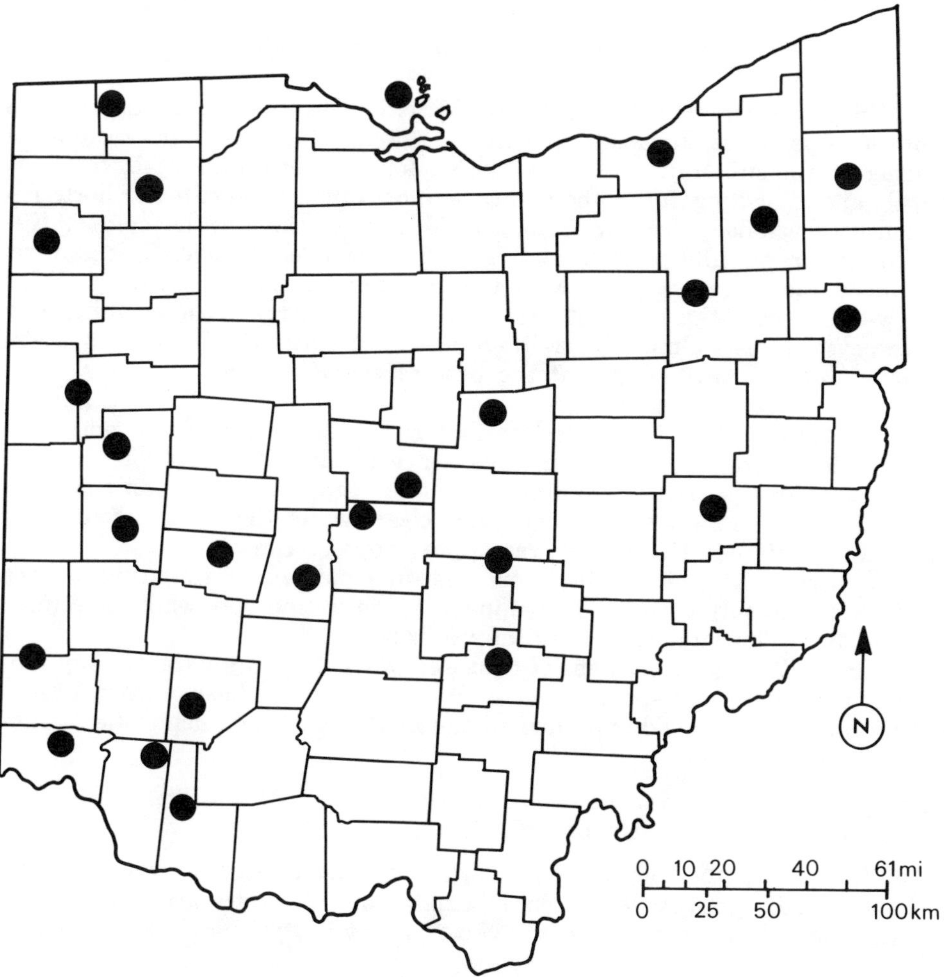

Figure 22. Ohio distribution of *Plumatella emarginata.*

flowing water better than do other species. On stone substrates in small Ohio rivers and streams, it is often the only phylactolaemate species present (Figure 22). On small concrete spillways draining enriched lakes and ponds, it sometimes grows as one continuous, thick, living blanket.

High tolerance to organic pollution also appears to be characteristic of *P. emarginata.* Colonies have been found in water contaminated by fecal wastes and domestic sewage, even in water described as "turbid and black" (Shrivastava and Rao, 1985). Bushnell (1974) reported colonies growing in the Kalamazoo River, Michigan, about 100 m downstream from the waste discharge of a chemical manufacturing plant.

Statoblasts in Ohio germinate in late April or early May, and the new colonies grow rapidly. Floatoblasts released in early June presumably germinate later in the same season (Table 7). Sessoblasts appear with increasing frequency towards September and October. Larva production has been described only from Cowan Lake in Clinton County, Ohio, where it occurs from early August to mid-September (Zimmerman, 1979).

Plumatella reticulata Wood, 1988

DESCRIPTION OF OHIO SPECIMENS. The colony has a neat and compact appearance. In small specimens, the tubular zooids are recumbant, attached to the substrate and to each other for most of their length. Increased growth and crowding force them into a more upright position. Short, free branches occasionally appear from the densest portion of the colony. Keel and furrow are usually prominent (Figure 23a). The ectocyst is dark and well sclerotized, but often covered with light-colored encrusting particles. When torn with forceps, the ectocyst feels brittle and gritty. With magnification, the ectocyst appears almost black in cross section under the layer of encrusted particles. Internal septa occur at the base of every branch.

The floatoblast is broad and symmetrical. Its straight sides and blunt ends give a distinctly rectangular appearance (Figures 23c-e). Float coverage is extensive, leaving only a small portion of the capsule exposed on the dorsal side. The dorsal fenestra is smooth and clear, while the ventral fenestra is lightly tuberculated. The two valves appear equally convex.

The sessoblast bears a distinctive net-like pattern of thin ridges across the frontal (dorsal) valve. These appear as dark lines on whole specimens (Figure 23b). The lamella is uniformly narrow.

Plumatella reticulata is easily confused with *P. emarginata.* Colony morphology is similar; both species have internal septa at the base of every branch, and both have a very wide annulus on the floatoblast. Key diagnostic features of *P. reticulata* include:

dark, net-like lines across the frontal valve of the sessoblast; these are unique among North American bryozoans, and;

in the floatoblast, the combination of straight sides, blunt ends, wide annulus, and equally convex valves; in *P. emarginata* the sides are more curved, the ends more tapered, and the valves strongly asymmetrical in side view.

TAXONOMY. For many years I considered *Plumatella reticulata* to be a form of *P. emarginata.* Side-by-side laboratory rearing of the two forms

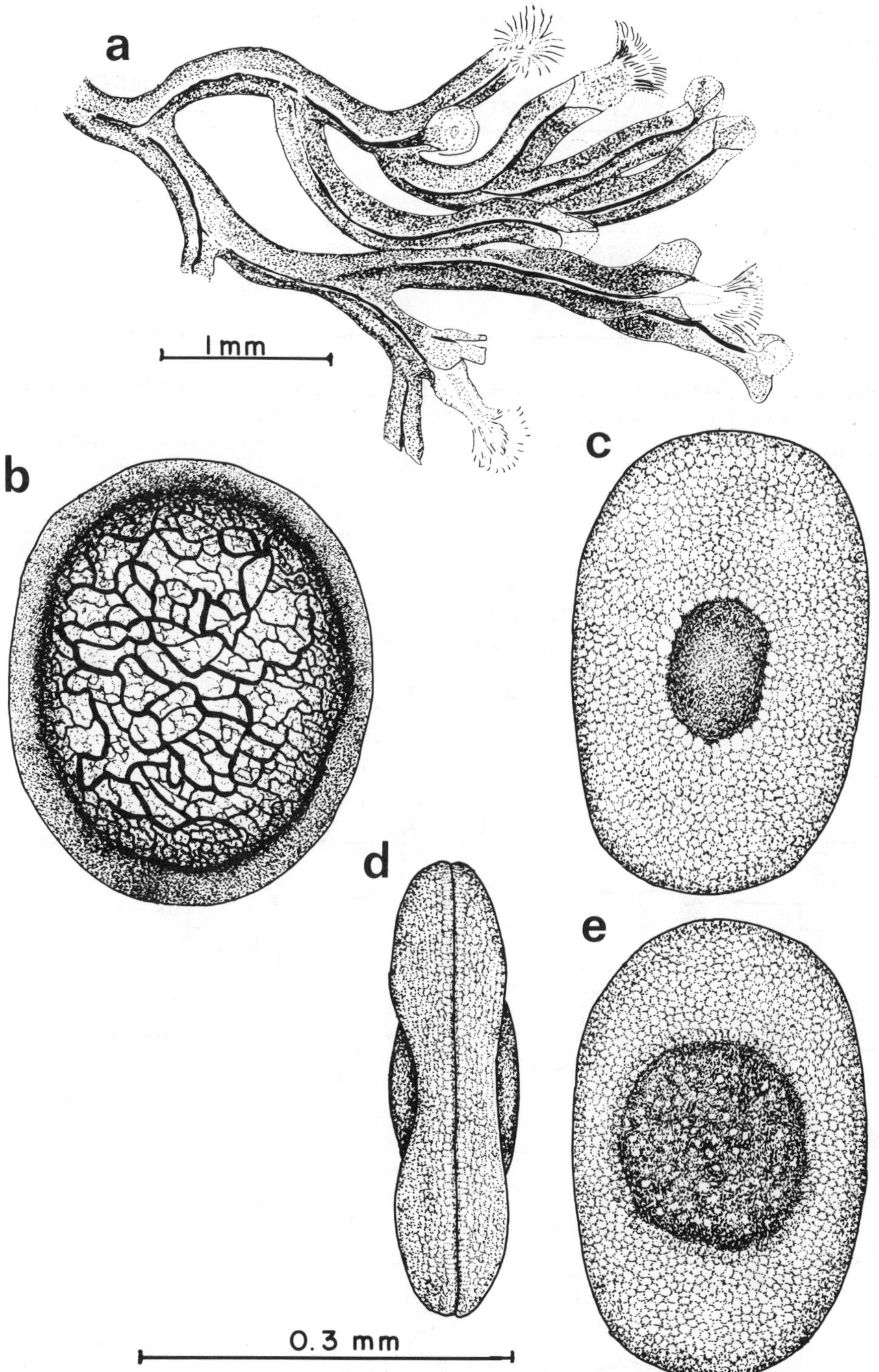

Figure 23. *Plumatella reticulata.* a—Portion of the colony, showing strong keel and furrow, and tendency of linear branches to fuse; b—e statoblasts: b—sessoblast, showing distinctive reticulated pattern; c—dorsal view of floatoblast; d—side view of floatoblast; e—ventral view of floatoblast.

Table 8. *Plumatella reticulata*. Measurements (dimensions in micrometers, standard deviations shown in parentheses.)

ITEM	MAXIMUM	MINIMUM	MEAN	NUMBER OF MEASUREMENTS
Number of tentacles	41	25	32 (2.6)	32
Floatoblast length	473	364	401 (22)	71
Floatoblast width	272	182	236 (13)	71
Floatoblast length/ width	2.1	1.5	1.7 (0.10)	71

finally showed that they belonged to distinct species (Wood, 1988). The same confusion may have been experienced by Hastings (1938) when she described *Plumatella emarginata,* Type 1 and Type 2, from a lake in what is now Israel.

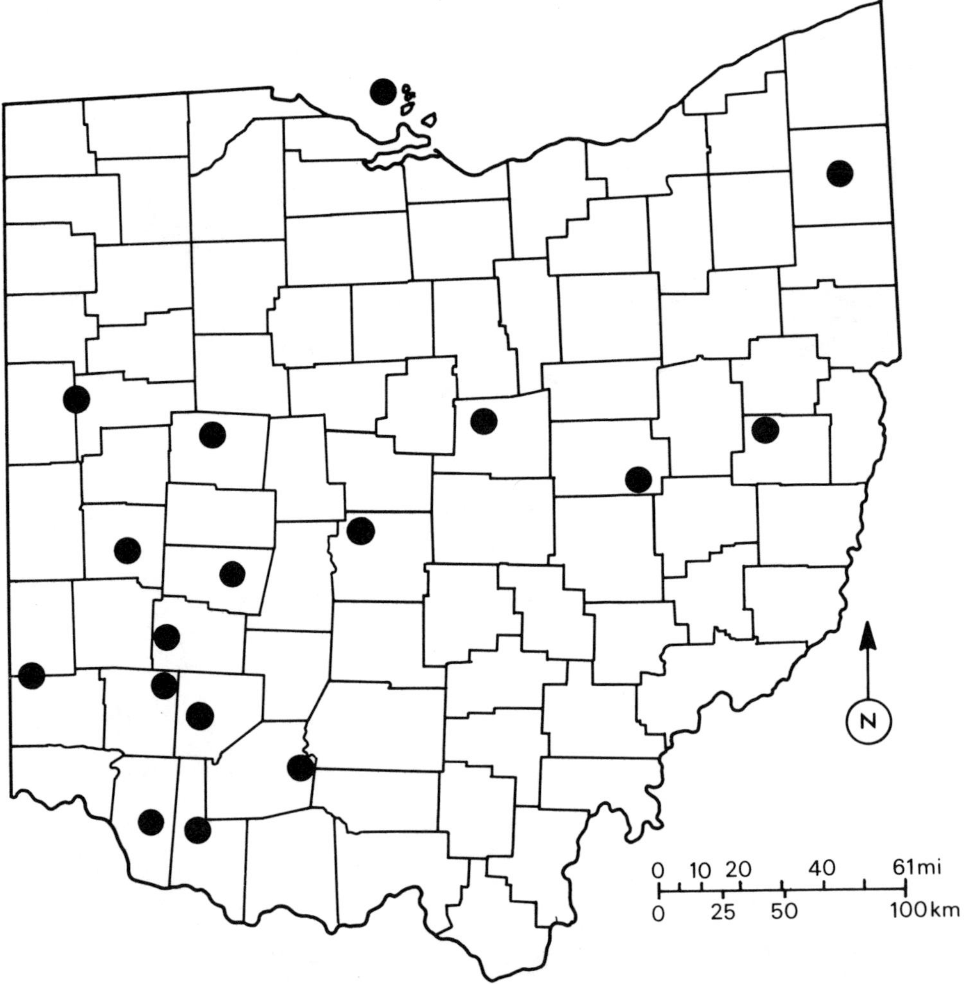

Figure 24. Ohio distribution of *Plumatella reticulata*.

Type 1 is clearly a typical *P. emarginata,* but the drawings and description of Type 2 floatoblasts exactly match floatoblasts of *P. reticulata* (Table 8). Unfortunately, the small amount of Type 2 material available to Hastings included no sessoblasts. Tentacle counts were not made.

BIOLOGY. *Plumatella reticulata* occurs over a wide range in Ohio (Figure 24). Fully developed colonies have been found from June through October, growing on wood and stone substrates in still water.

The lophophores of young zooids exhibit the same rhythmic flicking movement as in *Plumatella casmiana.* The beat begins shortly after the new zooid starts feeding, and it continues for up to 15 days. This behavior appears unrelated to feeding efficiency, but it may assist in the exchange of respiratory gases.

Hyalinella punctata Hancock, 1850

DESCRIPTION OF OHIO SPECIMENS. Young colonies consist of tubular strings of zooids adhering to the substrate throughout their length, with widely spaced and generally short branches, either straight or convoluted (Figure 25a). A colony may remain very open and linear, but tends to become dense and compact, with many upright, contiguous zooids, forming a uniform blanket over the substrate. Crowded zooids may be either separate or fused. The body wall is colorless and generally transparent, gelatin-soft to firm, elastic, and not easily torn, usually appearing thick and swollen. Emargination and keel are absent, and a barely perceptible furrow occurs only in rare instances when the ectocyst is lightly dusted with encrusting particles. In over 65% of Ohio specimens, the colony wall is dotted with minute white spots. Septa are absent, although a slight crease in the colony wall sometimes occurs between successive zooids.

There are apparently no sessoblasts in this species. Floatoblasts are the largest in North America lacking marginal processes (Table 9). Both the floatoblast and its capsule are broadly oval (Figures 25b-c). The annulus frequently is inflated only in irregular patches and is therefore not buoyant upon release. Separation of the valves reveals a distinct marginal serration on the dorsal half. The valves are equally convex. Both halves of the capsule are distinctly tuberculated.

Table 9. *Hyalinella punctata.* Measurements (dimensions in micrometers, standard deviations shown in parentheses).

ITEM	MAXIMUM	MINIMUM	MEAN	NUMBER OF MEASUREMENTS
Number of tentacles	50	34	41 (3.3)	102
Floatoblast length	765	489	538 (30.6)	206
Floatoblast width	408	255	357 (21.3)	206
Floatoblast length/ width	2.3	1.3	1.5 (1.10)	206

TAXONOMY. In 1850, Albany Hancock published a description of *Plumatella punctata* sp. nov. collected from two lakes in Northumberland, England. The features he noted and illustrated included a transparent ectocyst "freckled with minute opaque white spots," zooids appearing as "large conical cells tapering upwards towards the aperture, sometimes considerably and rather suddenly dilated at the base," and large, oval, black statoblasts.

Jullien (1885) considered *Plumatella punctata* synonymous with *P. repens* based on published descriptions. He did, however, establish the genus *Hyalinella* for *Plumatella*-like species having a gelatinous and mostly transparent colony wall, as distinguished from the sclerotized cuticle of *Plumatella* species. As the type species, Jullien chose *Plumatella vesicularis* Leidy, which he had not seen and for which only an incomplete description was available. Two years later, Kraepelin (1887) decided that *P. punctata* and *P. vesicularis* were synonymous, and these were combined under the name of *Hyalinella punctata*. A taxonomic study by Toriumi (1952b) and more recent observations of laboratory-reared colonies (Toriumi, 1972) have clarified much of the confusion over variable characters. Other detailed descriptions include those of Bushnell (1965c), Lacourt (1968), and Geimer and Massard (1986).

BIOLOGY. *Hyalinella punctata* has been collected in Ohio from May through October in lakes and ponds (Figure 26). Common substrates include submerged logs, boards, rocks, and the leaves of *Nelumbo lutea*. Elsewhere, colonies have been reported growing on *Nymphaea tuberosa* (Bushnell, 1966) and the submerged stems and fruiting heads of *Sagittaria* (Rogick, 1939). In Lake Erie, *Hyalinella punctata* occurs for the most part as small isolated

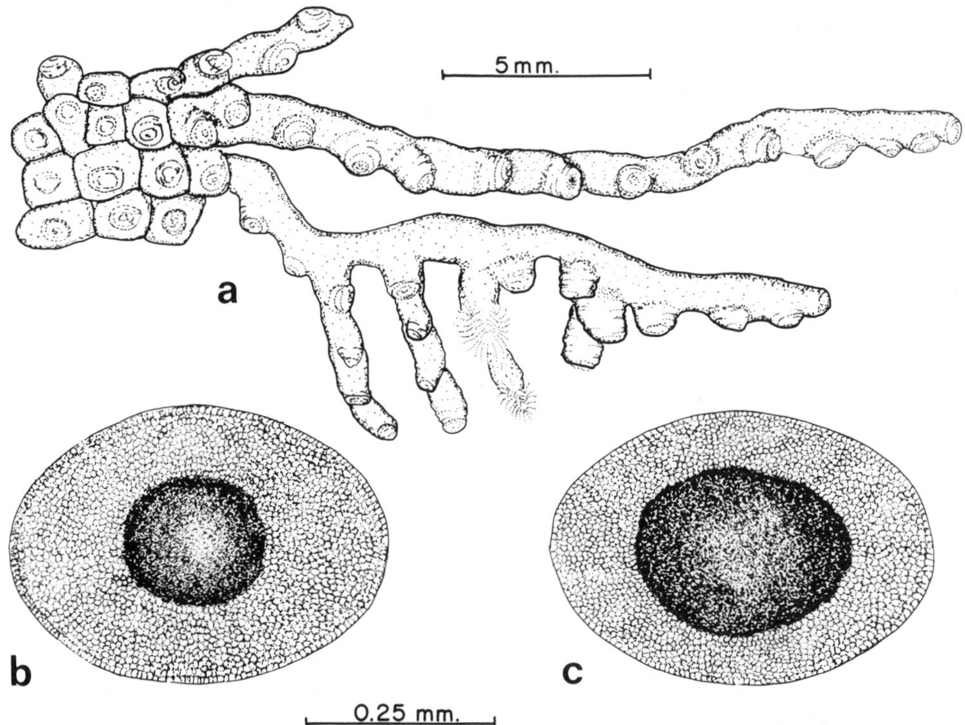

Figure 25. *Hyalinella punctata.* a—Colony fragment showing typical swollen appearance of zooid; b—dorsal floatoblast valve; c—ventral floatoblast valve.

44

patches on the undersides of rocks. On Middle Bass Island, however, a quiet pond 20 m inland contains several logs nearly covered with luxuriant *Hyalinella* growth.

Floatoblasts apparently are produced throughout the growing season, occurring in virtually all specimens collected from May through October. The lack of gas in the mature annulus is a curious but common feature which Toriumi (1972) has noticed also among his reared colonies. In some cases, the statoblast floats after being dried thoroughly, when water in the annulus is replaced by air.

The existence of sessile statoblasts in *Hyalinella punctata* was at first denied by Toriumi (1955b), but apparently later was allowed in a "special case" (Toriumi, 1972). Lacourt (1968) provides a description of "rarely found" sessoblasts, and Wiebach (1973) guardedly acknowledges their possible occurrence. However, no sessoblasts were found in the surveys by Annandale (1910) in Calcutta, Bushnell (1965c) in Michigan, Rogick (1935a) in Lake Erie, Toriumi (1941a; 1941b; 1942b) in Japan, Korea, and Taiwan, or in the present extensive survey of Ohio; nor were sessoblasts produced in any of 13 generations

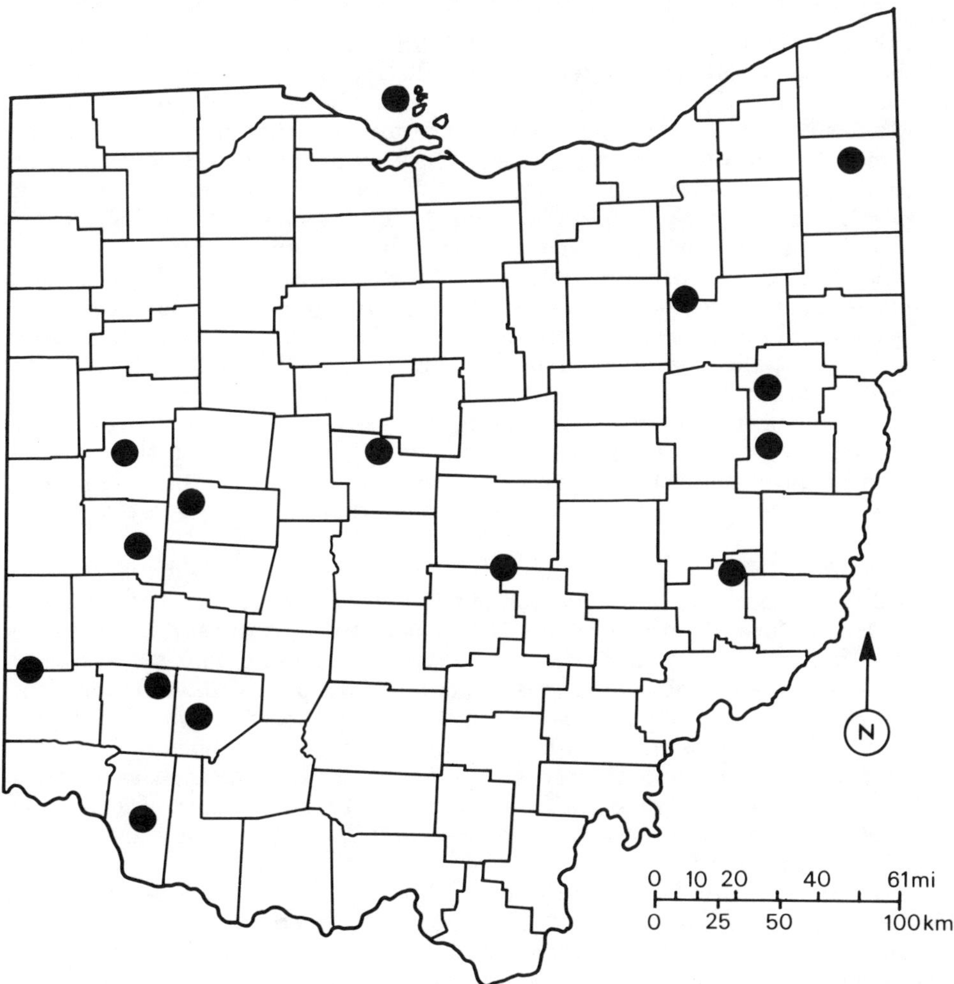

Figure 26. Ohio distribution of *Hyalinella punctata*.

45

of laboratory-reared colonies (Toriumi, 1972). Among the five other known species of *Hyalinella,* sessoblasts have been described for *H. minuta, H. indica,* and *H. vaihiriae.* Without convincing evidence in the case of *H. punctata,* however, the claim of sessoblasts in this species remains in doubt.

Dense colonies of *Hyalinella punctata* frequently have a blue-green appearance due to the growth of the cyanophyte, *Oscillatoria,* on the external colony wall. Trauberg (1940) reports a similar association between *Plumatella fungosa* and *Phormidium fragile.* The nature of this association has not been explored as yet.

Little is known of sexual reproduction in this species. Rogick (1939) reports larvae production in early August at Beechmont Lake, New Rochelle, New York, and she provides an excellent description of larval morphology and metamorphosis. In Ohio, larva release has been documented from early July through August (Zimmerman, 1979).

Family Lophophodidae

Lophopodella carteri Hyatt, 1868

DESCRIPTION OF OHIO SPECIMENS. The colony is small and unbranched, with a soft, transparent body wall that allows a clear view of the internal structures (Figure 27d). Usually having fewer than 20 zooids, the colony has a circular or oval appearance, generally less than 6 mm in diameter. There may be one or more small lobes. Fully extended zooids radiate in all directions from a central area. Under magnification, the outline of each living polypide has a distinctly irridescent quality. The dominant color of the gut varies with the type of food ingested, and has no taxonomic significance. Internally, each colony provides a common coelomic space shared by all zooids, without septa or partitions of any kind. The unpigmented mouth area distinguishes zooids of this species from those of *Pectinatella magnifica.* Among Ohio specimens, the number of tentacles ranges from 53 to 85, with a mean of 77.5 (50 counts, standard deviation=5.8).

Statoblasts are oval and curved along two axes, like a saddle (Figure 27a-c). An extensive annulus encroaches equally on both sides of the capsule. Each end of the statoblast has a series of small processes, each bearing minute barbs. The shape, dimensions, and numbers of these structures are variable.

TAXONOMY. When Carter described this species in 1859, he tentatively assigned it to the genus *Lophopus* (Dumortier, 1835) on the basis of its small, sac-like colony form. However, Hyatt (1868) believed that the statoblast spines suggested a close affinity to *Pectinatella magnifica,* and he named the species *Pectinatella carteri.* This name was acceptable to both Jullien (1885) and Davenport (1900). In 1904, Rousselet determined that for two reasons the statoblasts were sufficiently different to warrant a separate genus, *Lophopodella.* First, the spines are situated at the poles only, rather than around the entire periphery as in *Pectinatella.* Second, the barbs occur in linear series along both sides of the spines, whereas *Pectinatella* spines have only a single pair of terminal hooks.

Since 1904, a number of varieties of *Lophopodella* have been proposed (Annandale, 1911; Rogick, 1934; Borg, 1936), all based on variations of the statoblast. A second species, *Lophopodella pectinatelliformis* (Lacourt, 1959), known only from a single collection from Sumatra, has spines reduced to minute anchor-like projections.

BIOLOGY. *Lophopodella carteri* is distributed widely in Asia, occurring in India, Burma, China, Korea, Japan, and Java (Lacourt, 1968). Elsewhere, it has been reported from single collections in Kenya (De Beauchamp, 1936) and Australia (Colledge, 1917). In North America, the species is regionally rare but often locally abundant. Despite intensive search in Ohio, colonies could be found only in Lake Erie, where they sometimes nearly blanket rock and macrophyte substrates (Figure 28). Apparently, *L. carteri* was uncommon at the time of its discovery in Lake Erie in 1931, occurring mainly in a small

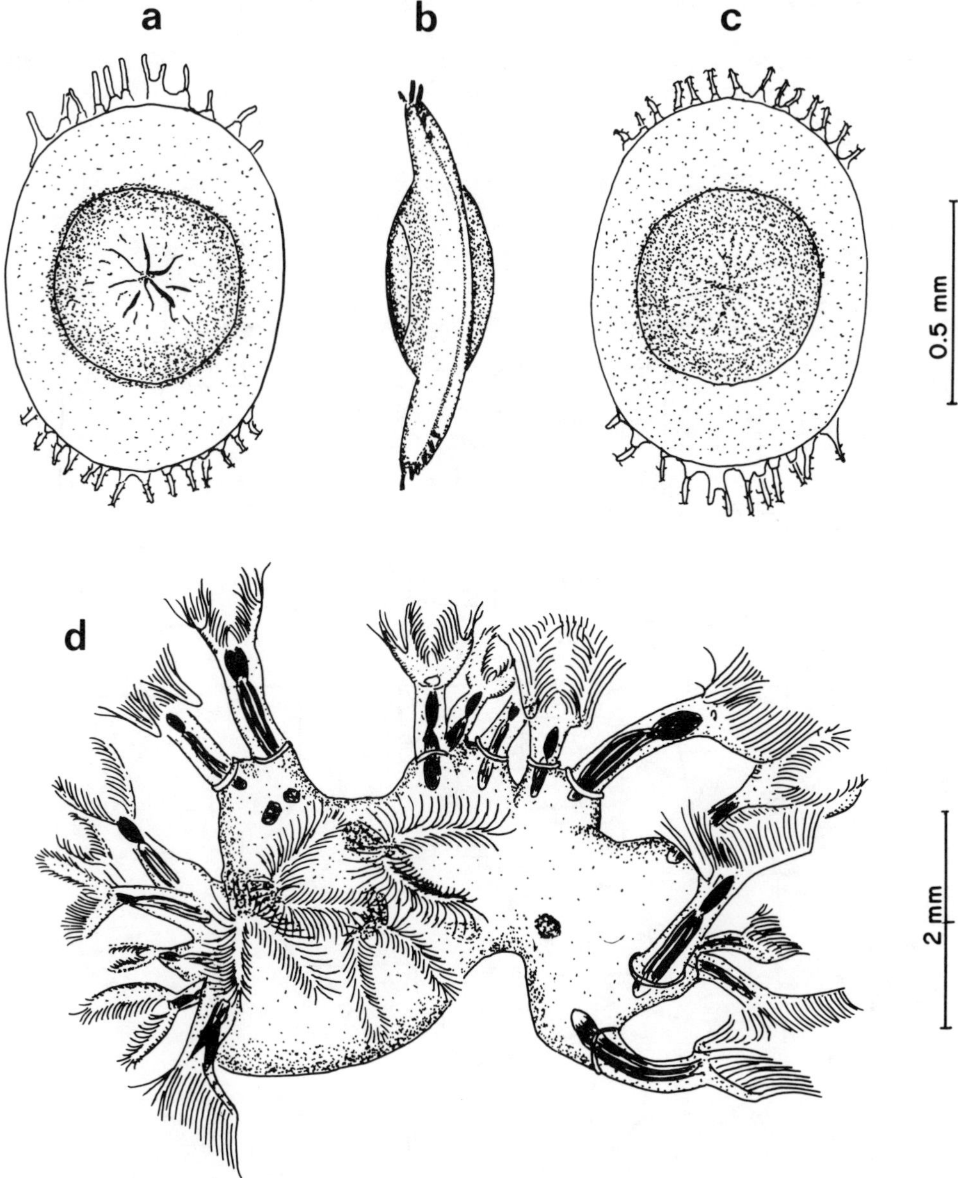

Figure 27. *Lophopodella carteri.* a—Floatoblast, dorsal view; b—floatoblast, side view; c—floatoblast, ventral view; d—colony form.

channel at East Harbor (Rogick, 1934 and unpublished notes). Tenney and Woolcott (1962) filed the first and only report of *L. carteri* from Virginia, where at a fish hatchery it "formed a continuous gelatinous layer on the walls and gate from the bottom of the spillway to within one inch of the water surface, covering a total area of 150 square feet." In Michigan, an extensive state-wide survey by Bushnell (1966) revealed *L. carteri* in only a single lake. Other occurrences in North America are noted in Pennsylvania and Kentucky (Rogick, 1957), Tennessee (Sinclair and Isom, 1963), New Jersey (Dahlgren, 1934), and Massachusetts (Smith, 1985).

Ecological data for this species are scant. Many submerged materials are used as substrates, including rocks and masonry, submerged wood, including creosote-treated oak pilings and live willow branches (Tenney and Woolcott, 1962), and such macrophytes as *Chara* (Bushnell, 1966), *Nymphaea, Vallisneria, Elodea, Potamogeton* (Rogick, 1934), and *Myriophyllum.* Active colonies have been collected in temperatures ranging from 8°C (Tenney and Woolcott, 1962) to 32°C (Bushnell, 1966). They thrive in both lotic and lentic habitats.

Figure 28. Ohio distribution of *Lophopodella carteri.*

Sexual reproduction was first noted by Hastings (1929) and fully described by Oka and Oda (1948) and by Oda (1960). Developing gonads were found in colonies collected near the Tone River, Japan, from late July through mid-September, and motile larvae were seen to emerge from these. In Ohio colonies, no sexual reproduction has been reported from natural populations. The period of greatest colony density and statoblast production is observed in late August through September.

Despite its spotty distribution, *Lophopodella carteri* has been studied extensively by various workers, probably owing to the particular ease of rearing and statoblast germination in the laboratory. Rogick has reported on colony development (1935b), histology (1937b), and statoblast dormancy (1940; 1941a). Mukai and Oda have examined statoblast morphology (1980a) and epidermal histology (1980b). Oda has described sexual reproduction and colony development (1960) as well as statoblast dormancy (1959), while Toriumi has examined intraspecific variation (1956; 1962; 1963a-d; 1964a-c; 1967a-c, and; 1974).

Variations in statoblast morphology were noted by Rogick (1936) for Lake Erie specimens and by Oda (1955) in Japan. Toriumi (1963a) demonstrated the effects of both temperature and diet on the number of spines of statoblasts in laboratory-reared colonies. He found further a correlation between the number of spines and the polar width of the statoblasts.

The toxicity of *Lophopodella carteri* colonies to fish was noted first by Rogick (1957). The poisonous substance apparently is released only when colonies are damaged or broken. In a study by Tenney and Woolcott (1964), a 0.09% filtered colony homogenate was lethal to three species of fish within one hour, and killed *Notropis analostanus* within 11 minutes. The toxin apparently is specific to fish. Even high concentrations have no apparent effect on white mice, tadpoles of *Rana pipiens*, or representatives of six tested phyla of common aquatic invertebrates.

Lophopodella carteri lends itself particularly well to experimentation and study. Compared to other phylactolaemate species, the zooids are remarkably insensitive to probing and vibration, and they are narcotized quickly by menthol. The transparent body wall allows detailed observations of internal movement and development. Colonies may be detached easily from one substrate and moved to another without damage. They grow well in the laboratory, dividing by fragmentation and producing abundant statoblasts. Statoblasts kept either wet or dry below 10°C remain viable for many months. Oda (1959) reported successful germination after cold storage for 4.5 years.

Pectinatella magnifica Leidy, 1851

DESCRIPTION OF OHIO SPECIMENS. The colony is gelatinous and firm, with a surface that is slimy to the touch. Young specimens may be small and globular (Figure 29d), similar to those of *Cristatella* and *Lophopodella*, later merging with other colonies and spreading as a thin layer over the substrate. Rapid growth produces a dense mat of zooids growing over a secreted mass of clear jelly. The surface appears divided into rosettes, each with 12-18 zooids. Massive colonies often exceed 60 cm in diameter (Figure 29e). The mouth region contains a conspicuous red pigment which distinguishes this from from all other gelatinous species in Ohio. A pair of milky-white spots occurs at the ends of the lophophore arms and on the anal side of the collar (Figure 29a). Tentacles range from 42 to 72 in Ohio specimens, with a mean of 56 (113 counts, standard deviation=6.9).

Statoblasts are circular, bent disks with 11-22 marginal spines, each bearing
a pair of distal hooks (Figure 29b-c). The statoblast has a buoyant annulus
with slightly greater capsule coverage on one side than the other.

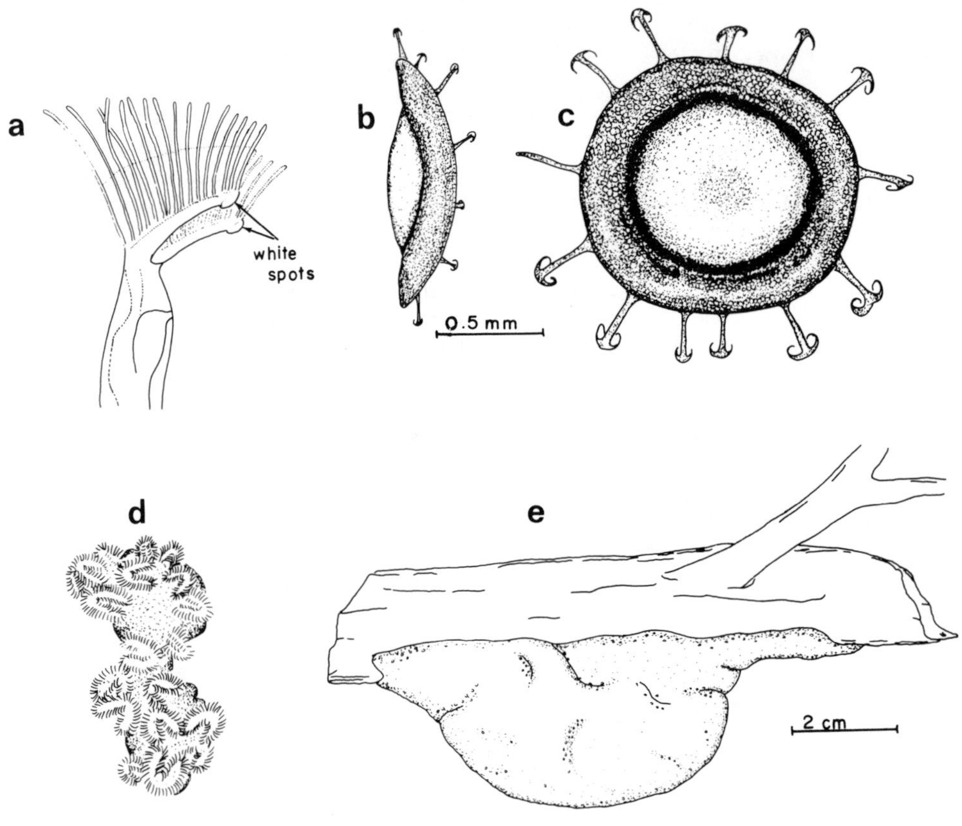

Figure 29. *Pectinatella magnifica.* a—Portion of a polypide showing prominent white spots on the
lophophore; b—side view of floatoblast; c—dorsal view of floatoblast; d—young colony; e—older
colony attached to the underside of a stick.

50

TAXONOMY. *Pectinatella magnifica* was described originally by Joseph Leidy in 1851 from specimens collected in the Philadelphia area. First believed to be a species of *Cristatella,* it later was given the name of *Pectinatella* by Kraepelin (1887). The only other species in this group, *Pectinatella gelatinosa,* now appears to be more closely related to *Lophopodella* (Mukai and Oda, 1980a).

BIOLOGY. *Pectinatella magnifica* occurs throughout eastern North America and central Europe, with single reports from Turkey and Guatamala. Recently it has become established in Japan, with compound colonies there reaching 3 m long and 50 cm wide (Oda, 1974). In Ohio, it is a common species, flourishing in large impoundments, and in sluggish rivers and backwaters (Figure 30). Typical substrates are submerged logs, twigs, and wooden dock pilings, on which the largest colonies generally occur below 1 m depth. Small specimens occasionally are found under large, loose rocks as well as on

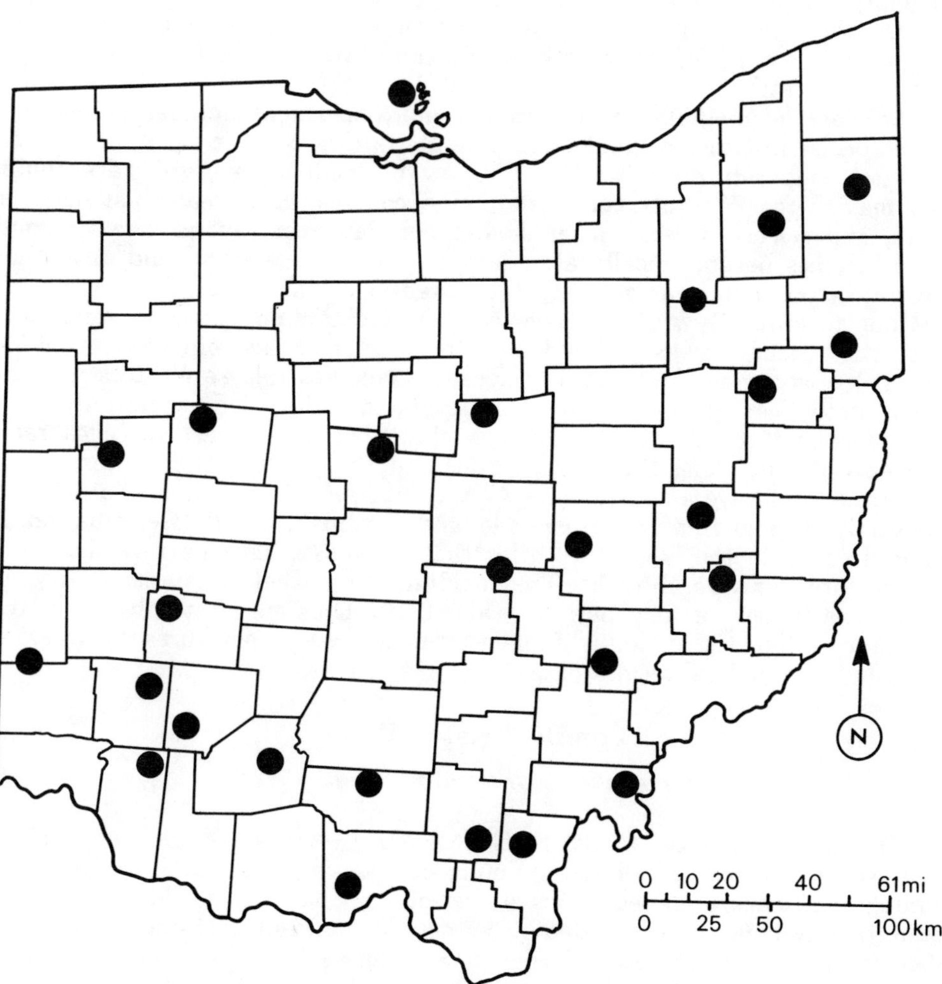

Figure 30. Ohio distribution of *Pectinatella magnifica.*

51

floating aquatic vegetation. Brown (1933) found colonies along stems of *Ceratophyllum* and Davenport (1904) reports colonies on "gates of dams" and on "shady structures, such as the south wall of reservoirs."

The massive size attained by *Pectinatella magnifica* makes this easily the most conspicuous of all freshwater bryozoan species in Ohio. The inner gelatinous mass, more than 99% water, also contains some sodium chloride and chitin, plus a high level of calcium and a protein chemically similar to egg albumin (Morse, 1930).

According to studies of Mukai and Oda (1980b), the epidermis of *Pectinatella magnifica* has a high proportion of vacuolar cells and bears an ectocyst only along the basal wall. The white spots have been shown to be organized masses of epidermal cells filled with eosinophilic granules of protein and lipid. Although Kraepelin (1887) referred to the white spots as epidermal glands, there is no evidence of a secretory function. The fact that they remain virtually intact after degeneration of the zooid suggests a possible role in waste storage.

Electrophoretic studies of three Ohio populations showed that only two of 17 presumptive genetic loci were polymorphic (Ingold et al, 1984). Percent polymorphism (5.9%) and mean heterozygosity (0.013%) were surprisingly low. On the other hand, a sharp and consistent difference at one locus was seen between a population in northern Ohio and two others in the southwestern part of the state.

Pectinatella magnifica is apparently more thermophilic than other bryozoan species in Ohio. Statoblasts germinate during May or early June when the water temperature reaches 18°-20°C. As the small colonies grow, eventually they may merge. By July these compound colonies are already fist-sized or larger, and release several dozen larvae per day over several weeks. Small larval colonies become locally abundant on sticks, vegetation, and any other substrate. It is not clear whether the parental colonies then disintegrate or continue to grow. By August, large colonies are common in Ohio reservoirs. Often they become dislodged and float to the surface, accumulating along lee shores. In September and October decaying colonies release clusters of individual statoblasts encased in membrane and a watery jelly. The jelly soon disappears and the floating statoblasts cling together in ragged, black rafts in open water or along the shore, throughout the autumn and winter.

Statoblasts have a suture zone with a single, mucilaginous medial rib and oppositely arranged lateral ribs (Rao and Bushnell, 1979). Germination is induced easily in the laboratory after the statoblasts have been held at cool temperatures for several weeks. They exhibit only a weak resistance to drying, but can be frozen for long periods (Oda, 1979). Clusters of statoblasts in the laboratory often are killed by fungus even at low temperatures, so freezing in ice is best for long-term storage.

Family Cristatellidae

Cristatella mucedo Cuvier, 1798

DESCRIPTION OF OHIO SPECIMENS. Colonies are clear and gelatinous, without lobes or branches. Young colonies are circular in outline, becoming briefly heart-shaped and then distinctively elongate as older specimens. Ohio colonies have been recorded at 5-10 cm long and up to 5 cm wide (Figure 31a). Bushnell (1965a) reports colonies as long as 35 cm and 50 cm from a lake in southwestern Michigan. There are no white spots or pigmentation on the zooids or colony wall. The gelatinous ectocyst is very thin except along the flat colony base, where it can be several millimeters thick.

Statoblasts are circular, never bent, with long, wiry spines extending outward from the capsule beyond the periphery (Figure 31b-d). Each spine bears a forked point with up to 12 small hooks. The spines are longer and more numerous on the ventral valve than on the dorsal. The annulus arises on the dorsal side and extends around to the ventral valve as an unattached, membranous collar, concealing the suture area of the capsule. All statoblasts are buoyant upon release.

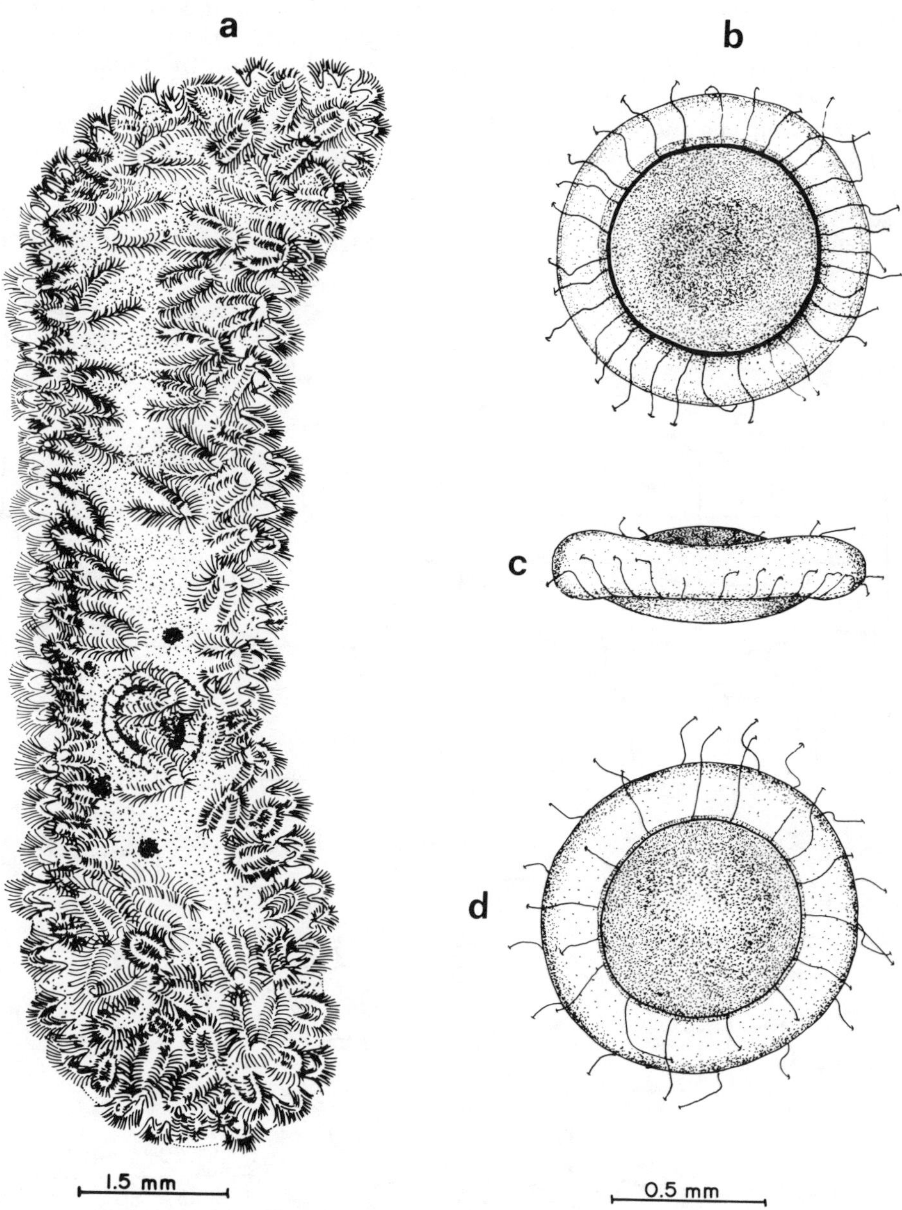

Figure 31. *Cristatella mucedo*. a—Typical colony of moderate size; b—dorsal view of floatoblast; c—side view of floatoblast; d—ventral view of floatoblast.

TAXONOMY. *Cristatella* was the first genus to be established among the freshwater bryozoa, with *Cristatella mucedo* as its single member. Several other species that have been proposed on the basis of somewhat variable features generally have not been accepted.

BIOLOGY. *Cristatella mucedo* has been reported throughout North America, Europe, Western Asia, and Japan. Habitats range from snow-fed alpine lakes to warm, eutrophic ponds, even including acid peat lakes of northern Germany. The population size varies from year to year, and even over a single season. In Ohio, the species was found only in Lake Erie and in Punderson Lake (Geauga County), although it may occur elsewhere in the state (Figure 32).

Cristatella is one of the few bryozoan species that can be found on top of a substrate, "delighting in sunlight," as Allman (1856) put it. In Lake Erie there are conspicuous colonies on the upper surfaces of large rocks, sometimes resembling giant fuzzy caterpillers. Aquatic vegetation also serves commonly as a substrate. Bushnell (1966) reports Michigan specimens clinging to stems

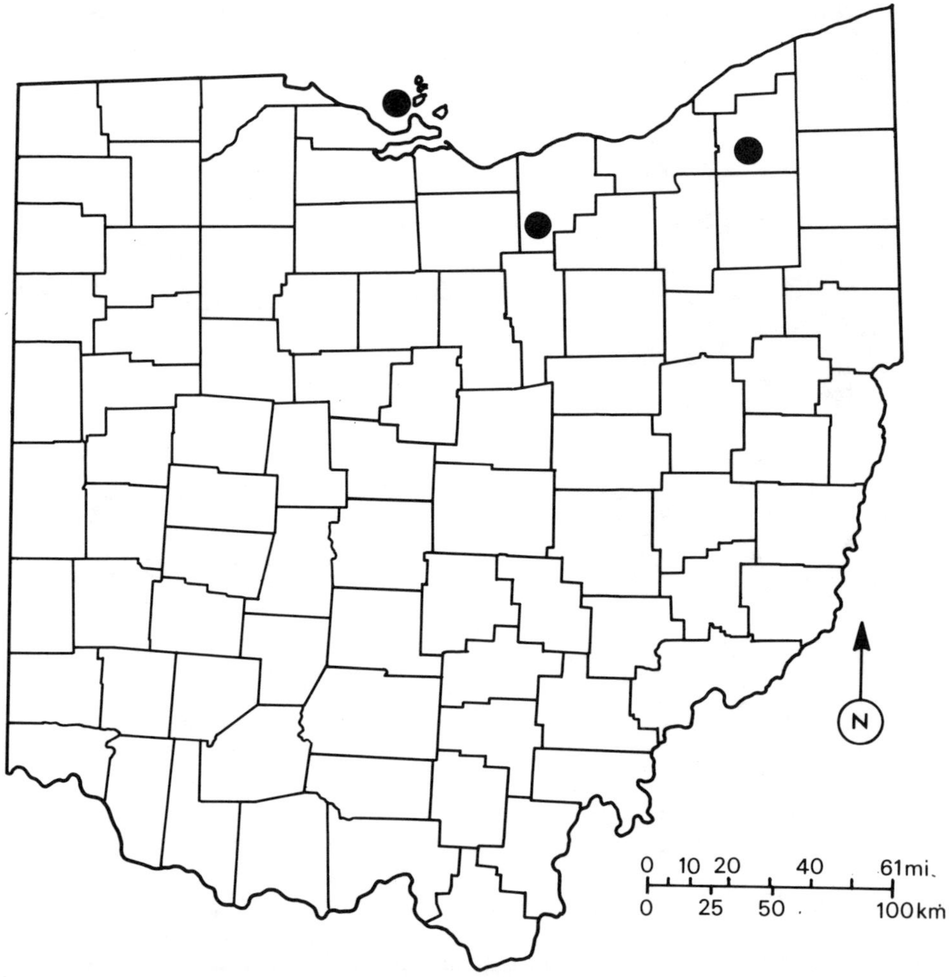

Figure 32. Ohio distribution of *Cristatella mucedo.*

of *Ceratophyllum, Chara, Nuphar, Nymphaea, Pontederia, Ranunculus,* and *Typha.*

According to Kraepelin (1887), statoblasts in Germany germinate in May and June, and colonies grow slowly for the first month or so. Greatest growth occurs in July and August, when colonies reproduce by fragmentation and presumably also by the sexual production of larvae. These larvae measure about 1 mm in diameter and bear two to five zooids each (Toriumi, 1941a). New statoblasts appear in August, but do not germinate until the following year.

Statoblasts are very resistant to dessication and freezing. It is common to discover old valves in a lake or pond without ever finding a colony. Subfossil statoblasts have been found in Canadian sapropel and peat deposits dating from the late Tertiary or early Pleistocene (Kuc, 1973).

CLASS GYMNOLAEMATA

ORDER CTENOSTOMATA

Family Paludicellidae

Paludicella articulata (Ehrenberg, 1831)

DESCRIPTION OF OHIO SPECIMENS. The colony is tubular and sparsely branched, with both free and attached components (Figure 33a). Each branch is a series of very slender, spindle-shaped zooids which are joined end-to-end and separated from each other by complete internal septa. Lateral branches occur almost at right angles, either singly or in opposite pairs. The colony wall is rigid and clear with little or no incrustation. The lophophore projects obliquely from the zooid chamber through a square orifice (Figure 33b). Tentacles are short, few in number, and arranged in a circle around a central mouth (Figure 33c). Instead of statoblasts, there are other overwintering structures termed hibernacula, which are irregularly shaped bodies somewhat smaller than the zooids. They are formed in various locations as structural members of the colony.

TAXONOMY. Although the name has been changed several times since Ehrenberg's description, first published in 1831, the taxonomic status of *Paludicella articulata* never has been questioned seriously. The genus also includes *Paludicella pentagonalis,* known so far only from Latin America, in which species the orifice is reported to have five sides rather than four.

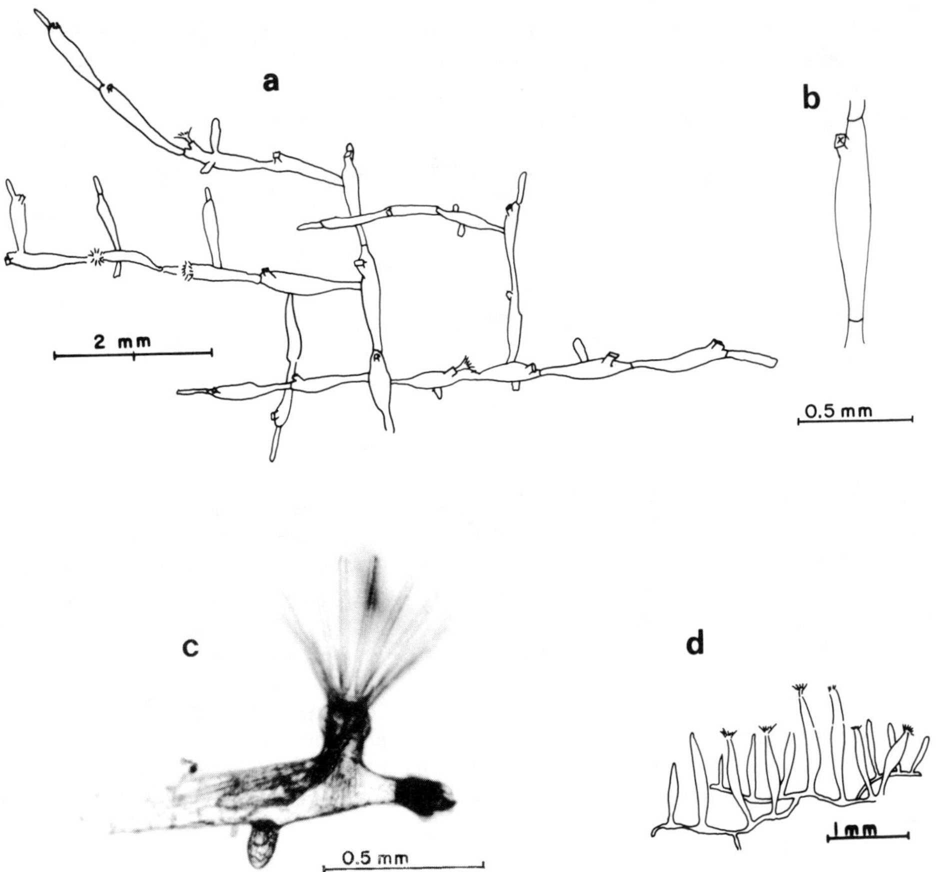

Figure 33. Freshwater gymnolaemates. a—Colony of *Paludicella articulata;* b—zooid of *Paludicella articulata* showing four-sided orifice; c—budding zooid of *Paludicella articulata;* d—colony of *Pottsiella erecta* showing stolons between the zooids.

BIOLOGY. *Paludicella articulata* is found on wood, stone, and other submerged substrates, including living plants. It grows well in both calm and rapidly flowing water, but where wave action is strong it is restricted to the protected sides and crevices of rocks.

Toriumi (1952a) noted that colonies in Japan varied widely in zooid shape, density, and the point of origin of the branches. Occasionally, several types were found growing at the same time in different parts of a single pond. When colonies were brought into the laboratory, short zooids produced long ones and branching was uniform, suggesting an important influence of environmental factors (Table 10).

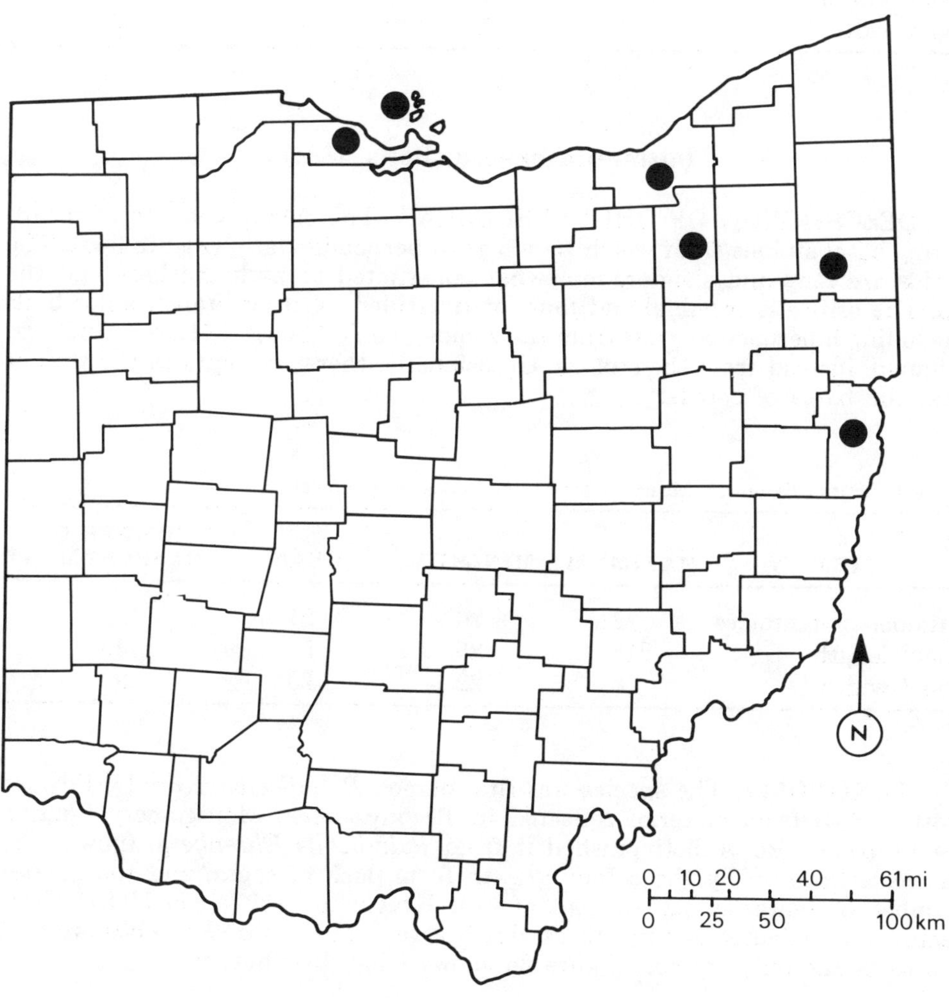

Figure 34. Ohio distribution of *Paludicella articulata*.

Paludicella articulata is one of the most widely distributed freshwater bryozoan species, having been reported from Europe, India, China, and New Zealand. It has been abundant in Lake Erie at least since 1932, but surprisingly was not mentioned there in reports by Davenport (1904) or Landacre (1901) (Figure 34). Elsewhere in Ohio, the species has been found recently only in a few northeastern sites. However, Bushnell (1965a) reported it to be the third most frequently encountered species throughout Michigan. Sinclair and Isom (1963) found it in many locations in the Tennessee River.

Table 10. *Paludicella articulata.* Measurements (dimensions in micrometers).

ITEM	MAXIMUM	MINIMUM	MEAN	NUMBER OF MEASUREMENTS
Number of tentacles	23	14	17	68
Zooid length	140	60	90	45
Zooid width	30	19	—	45

Pottsiella erecta (Potts, 1884)

DESCRIPTION OF OHIO SPECIMENS. The colony consists of meandering basel stolons from which zooids grow perpendicularly (Figure 33d). The zooids are long and slender, somewhat constricted at both the base and the tip. The orifice is terminal and four- or five-sided. Neither branches nor buds (including hibernacula) grow from the zooids (Table 11). All of these structures originate instead from the stolons. Occasionally, there are septa in the stolons near the bases of zooids.

Table 11. *Pottsiella erecta.* Measurements (dimensions in micrometers).

ITEM	MAXIMUM	MINIMUM	MEAN	NUMBER OF MEASUREMENTS
Number of tentacles	22	20	21	3
Zooid length	218	85	155	30
Zooid width	28	22	26	30

TAXONOMY. The species was first named *Paludicella erecta* by Edward Potts (1884) from material collected in Tacony Creek, Montgomery County, Pennsylvania. Potts distinguished it from *Paludicella Ehrenbergi* (now *Paludicella articulata*) by the colony growth form, lack of septa, and the greater number of tentacles (20 instead of 16). Kraepelin (1887) considered these features sufficiently distinct to establish a separate genus, *Pottsiella* with *P. erecta* as the only species. Figure 35 shows Ohio distribution.

BIOLOGY. Unlike most other species of bryozoans, *Pottsiella erecta* occurs frequently on the upper surface of substrate, which includes both stone and wood. In Lake Erie material in my custody, collected by Jerry H. Hubschman, the stolons were growing within soft, rotted wood, with zooids protruding at intervals.

Potts found his Pennsylvania colonies to be associated frequently with the sponge, *Trochospongilla leidyi.* In Lake Erie, Maciorowski (1974) collected *Pottsiella erecta* in close association with organisms which included the sponge, *Heteromyenia latitentus,* a frustule of *Craspedacusta sowerbyi,* numerous unidentified sessile rotifers, the entoproct *Urnatella gracilis,* and six other bryozoan species

Reports of *Pottsiella erecta* are, to date, too sparse to reveal a pattern in habitat preference. Lake Erie specimens have been found so far only at depths greater than 5 m, but downstream from the Mahoning Reservoir it grew only a few inches beneath the surface in rapidly flowing water. Everitt (1975) reports colonies to be often abundant in rivers, creeks, and borrow pits of southwestern Louisana.

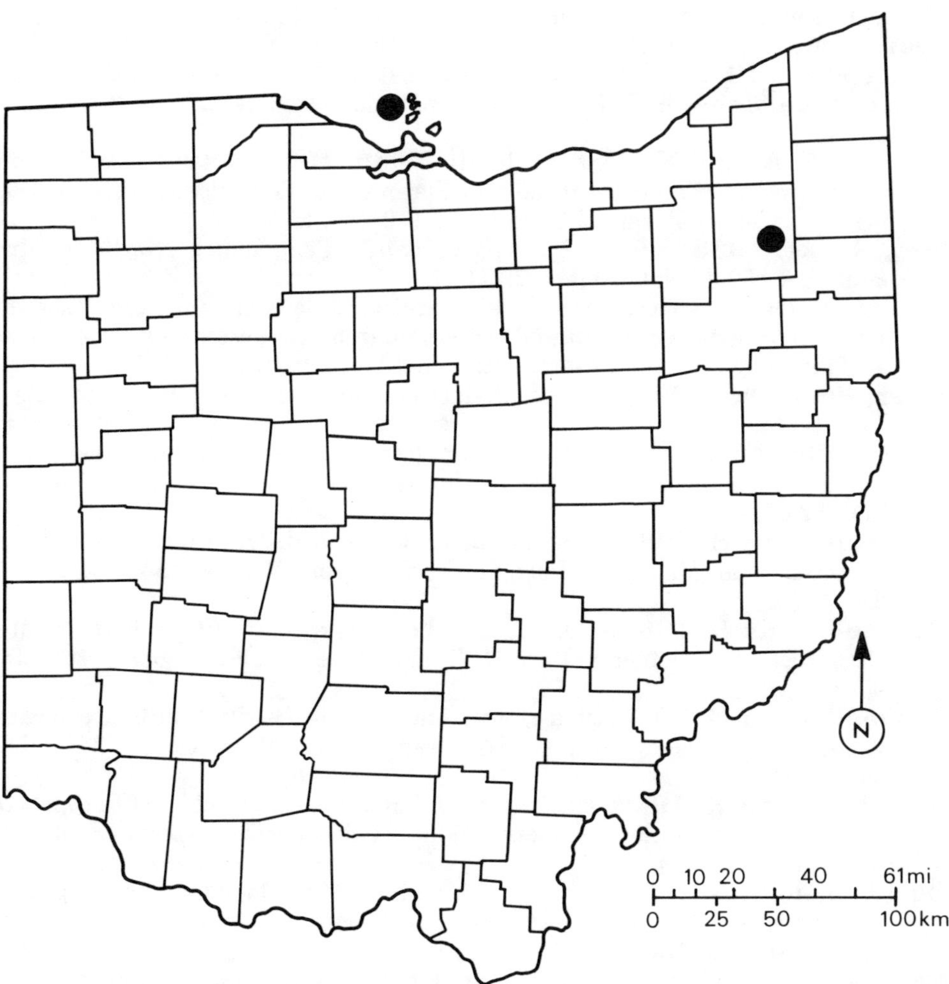

Figure 35. Ohio distribution of *Pottsiella erecta.*

LITERATURE CITED

Allman, George J. 1844. On *Plumatella repens*. Rep. Brit. Assn. for the Advancement of Science 13(2):74-76.

Allman, George J. 1856. A monograph of the fresh-water Polyzoa. Ray Society 28:1-119.

Anderson, Dennis M. 1983. The natural divisions of Ohio. Nat. Areas J. 3(2):23-33.

Annandale, Nelson. 1909. A new species of *Fredericella* from the Indian lakes. Rec. Ind. Mus. 3:373-374.

Annandale, Nelson. 1910. Materials for a revision of the phylactolaematous polyzoa of India. Rec. Ind. Mus. 5:37-57.

Annandale, Nelson. 1911. Fresh-water sponges, hydroids, and polyzoa. Fauna of British India, London, 2 (Bryozoa):161-251.

Backus, Bryon and Timothy S. Wood. 1981. Karyotypic and morphological confirmation of species in *Fredericella australiensis* (Bryozoa: Phylactolaemata). Trans. Amer. Microsc. Soc. 100(3):253-263.

Beauchamp, P. de 1936. Tubelliares et Bryozoaires. Mission scientifique de l'Omo, III. Zoologie. Mém. Mus. Hist. Nat. Paris 4:141-153.

Blumenbach, Johann F. 1779. Handbuch der Naturgeschichte. Göttingen. 117 p.

Bonetto, A. A. and E. Cordiviola de Yuan. 1965. Notas sobre briozoos (Endoprocta y Ectoprocta) del Rio Parana. III. *Fredericella sultana* (Blumenbach) en el Parana Medio. Physis 25(70):255-262.

Borg, Folke. 1936. Sur quelques Bryozoaires d'eau douce Nord-Africains. Bull. Soc. Hist. Nat. Alger 27:271-283.

Braem, F. 1911. Bryozoa und deren Parasiten. Beiträge zur Kenntniss der Fauna Turkestans auf Grund des von Pedaschenko gesammelten Materials 7. Trav Soc. nat. St. Petersbourg (Zool.) 42(2):1-56.

Brien, Paul. 1953. Etude sur les phylactolémates. Ann. Soc. Roy. Zool. Belg. 84:301-444.

Brown, Claudeus J. D. 1933. A limnological study of certain fresh-water Polyzoa with special reference to their statoblasts. Trans. Amer. Microsc. Soc. 52:271-314.

Bushnell, John H. 1965a. On the taxonomy and distribution of freshwater Ectoprocta in Michigan. Part I. Trans. Amer. Microsc. Soc. 84(2):231-244.

Bushnell, John H. 1965b. On the taxonomy and distribution of freshwater Ectoprocta in Michigan. Part II. Trans. Amer. Microsc. Soc. 84(3):339-358.

Bushnell, John H. 1965c. On the taxonomy and distribution of freshwater Ectoprocta in Michigan. Part III. Trans. Amer. Microsc. Soc. 84(4):529-548.

Bushnell, John H. 1966. Environmental relations of Michigan Ectoprocta, and dynamics of natural populations of *Plumatella repens*. Ecological Monogr. 36:95-123.

Bushnell, John H. 1968. Aspects of architecture, ecology, and zoogeography of freshwater Ectoprocta. Atti. Soc. Ital. Sci. Nat. Mus. Civ. Stor. Nat. Milano 108:129-151.

Bushnell, John H. 1971. Porifera and Ectoprocta in Mexico: architecture and environment of *Carterius latitentus* (Spongillidae) and *Fredericella australiensis* (Fredericellidae). Southwest Nat. 15:331-346.

Bushnell, John H. 1974. Bryozoans (Ectoprocta). Pages 157-194, *in* Pollution Ecology of Freshwater Invertebrates. Charles W. Hart, Jr. and Samuel L. H. Fuller, eds. Academic Press. New York/London. 389 p.

Carter, Henry J. 1859. On the identity in structure and composition of the so-called seed-like body of *Spongilla* with the winter-egg of the Bryozoa; and the presence of starch granules in each. Ann. Mag. Nat. Hist. (3) 3:331-343.

Colledge, W. R. 1917. *Lophopus brisbanensis* sp. nov. Proc. Roy. Soc. Queensland 29:123-124.

Cuvier, Georges. 1798. Tableau élémentaire de l'histoire naturelle des animaux. 710 p.

Dahlgren, Ulric. 1934. A species and genus of freshwater Bryozoa new to North America. Science 79(2057):510.

Davenport, Charles B. 1890. *Cristatella:* the origin and development of the individual in the colony. Bull. Mus. Comp. Zool. Harvard College 20:101-151.

Davenport, Charles B. 1900. On the variation of the statoblasts of *Pectinatella magnifica* from Lake Michigan at Chicago. Amer. Nat. 34:959-968.

Davenport, Charles B. 1904. Report on the fresh-water Bryozoa of the United States. Proc. U.S. Nat. Mus. 27:211-221.

Dumortier, Barthélemy C. 1835. Recherches sur l'anatomie et la physiologie des Polypiers composés d'eau douce. Bull. Acad. Roy. Sci. Bruxelles 2:421-454.

Dumortier, Barthélemy Charles and Pierre Joseph van Beneden. 1850. Histoire naturelle des polypes d'eau douce. IIe partie. Descriptions. Mémoire servant de complément au tome XVI des Mémoires de l'Académie royale des Sciences et Belles-lettres de Bruxelles. Mém. Acad. Sci. Bruxelles 16 (complément):1-96.

Ehrenberg, Christian G. 1831. Phytozoa polypi. Symbolae physicaem etc. Animalia Evertebrata (1):831.

Everitt, Betty. 1975. Freshwater Ectoprocta: distribution and ecology of five species in southeastern Louisiana. Trans. Amer. Microsc. Soc. 94:130-134.

Franzén, Åke. 1982. Ultrastructure of spermatids and spermatozoa in the fresh water bryozoan *Plumatella* (Bryozoa, Phylactolaemata). J. Submicrosc. Cytol. 14:323-336.

Franzén, Åke and Terry Sensenbaugh. 1983. Fine structure of the apical plate in the larva of the freshwater bryozoan *Plumatella fungosa* (Pallas) (Bryozoa: Phylactolaemata). Zoomorphology 102:87-98.

Geimer, Gaby and Jos. A. Massard. 1986. Les Bryozoaires du Grand-Duché de Luxembourg et des régions limitrophes. Mus. Hist. Nat. Luxembourg. 188 p.

Gervais, Paul. 1838. Sur les polypes d'eau douce. Extraits Proces-Verbaux des Séances de la Société Philomatique 1838:129, 130.

Goddard, Edward J. 1909. Australian freshwater Polyzoa. Part I. Proc. Linn. Soc. New S. Wales 34:487-496.

Gruncharova, Tyana. 1971. *Internectella bulgarica* nov. gen., nov. sp. A new bryozoan species (Bryozoa, Phylactolaemata). Compt. rend. Acad. Bulg. Sci. 24(3):361-364.

Hancock, Albany. 1850. On the anatomy of the fresh-water Bryozoa, with descriptions of three new species. Ann. Mag. Nat. Hist. (2) 5:175-202.

Hastings, Anna B. 1929. Notes on some little-known phylactolaematous Polyzoa and description of a new species from Tahiti. Ann. Mag. Nat. Hist. 10(3):300-310.

Hastings, Anna B. 1938. The Polyzoa. Pages 529-535, *in* Report of the Percy Sladen Expedition to Lake Huleh: A Contribution to the Freshwaters of Palestine. R. Washbourne and R. F. Jones, eds. Ann. Mag. Nat. Hist. Number 11.

Hôzawa, Sanji and Makoto Toriumi. 1940. Some freshwater Bryozoa found in Manchoukuo. Rep. Limnobiol. Surv. Kwantung Manchoukuo 3:425-434.

Hubschman, Jerry H. 1970. Substrate discrimination in *Pectinatella magnifica* Leidy (Bryozoa). J. Exp. Biol. 52:603-607.

Hyatt, Alpheus. 1866. Observations on polyzoan suborder Phylactolaemata. Comm. Essex Inst. 4:107-228.

Hyatt, Alpheus. 1868. Observations on polyzoan suborder Phylactolaemata. Comm. Essex Inst. 5:97-112; 145-160; 193-232.

Ingold, James L., Neal D. Mundahl, and Sheldon I. Guttman. 1984. Ecology and population genetics of the freshwater bryozoan, *Pectinatella magnifica* Leidy. J. Freshwater Ecology 2(5):499-508.

Job, Pierre. 1976. Intervention des populations de *Plumatella fungosa* (Pallas) (Bryozoaire phylactolème) dans l'autoépuration des eaux d'un étang et d'un ruisseau. Hydrobiologia 48(3): 257-261.

Jónasson, Pétur M. 1963. The growth of *Plumatella repens* and *P. fungosa* (Bryozoa Ectoprocta) in relation to external factors in Danish eutrophic lakes. Oikos 14(2): 121-137.

Jullien, J. 1885. Monographie des Bryozoaires d'eau douce. Bull. Soc. Zool. France 10:91-207.

Kent, William S. 1870. On a new polyzoan, *Victorella pavida,* from the Victoria docks. Quart. J. Microsc. Sci. n.s. 10:34-39.

Kraepelin, Karl. 1887. Die deutschen Süswasser-Bryozoen. Eine Monographie. Abh. Naturwiss. Verein Hamburg 10. 168 p.

Kuc, M. 1973. Fossil statoblasts of *Cristatella mucedo* Cuvier in the Beaufort Formation and in interglacial and postglacial deposits of the Canadian arctic. Geological Survey of Canada Paper 72-28, Ottawa. 12 p.

Lacourt, A. W. 1959. *Lophopodella pectinatelliformis* nov. sp. (Bryozoa-Phylactolaemata). Zool. Meded. 36(17):273-274.

Lacourt, A. W. 1968. A monograph of the freshwater Bryozoa-Phylactolaemata. Zool. Verhandel., No. 93. Rijksmuseum van Nat. Hist. Leiden. 159 p.

Landacre, F. L. 1901. Sponges and bryozoans of Sandusky Bay. Ohio Nat. 1:96-97.

Leidy, Joseph. 1851. On *Cristatella magnifica* n. sp. Proc. Acad. Nat. Sci. Philadelphia 5:265-266.

Linnaeus, Carolus. 1758. Systema naturae. 10 ed.

Maciorowski, Anthony F. 1974. The occurrence of the freshwater bryozoan *Pottsiella erecta* (Potts) 1884 (Gymnolaemata: Paludicellidae) in Lake Erie. Ohio J. Sci. 74(4):245-247.

Marcus, Ernst. 1934. Über *Lophopus crystallinus* (Pall.) Zool. Jahrb. Abt. Anat. Ontog. Tiere 58:501-606.

Marcus, Ernst. 1941. Sôbre Bryozoa do Brasil. Bol. Fac. Fil. Ciênc. Letr. Sâo Paulo Zoo. 5:3-208.

Marcus, Eveline DuBois-Reymond. 1946. On a new Brazilian form of *Fredericella sultana* (Bryozoa, Phylactolaemata). Com. Zool. Mus. Hist. Nat. Montevideo 2:1-10.

Marcus, Eveline DuBois-Reymond. 1953. Bryozoa from Lake Titicaca. Bol. Fac. Ciênc. Letr. Sâo Paula (Zool.) 18:149-163.

Morse, Withrow. 1930. The chemical constitution of *Pectinatella*. Science 71:265.

Mukai, Hideo and Shuzitu Oda. 1980a. Comparative studies on the statoblasts of higher phylactolaemate bryozoans. J. Morphology 165:131-155.

Mukai, Hideo and Shuzitu Oda. 1980b. Histological and histochemical studies on the epidermal system of higher phylactolaemate bryozoans. Annot. Zool. Jap. 53:1-17.

Mukai, Hideo, Masashi Tsuchiya, and Kazuki Kimoto. 1984. Fusion of ancestrulae germinated from statoblasts in plumatellid freshwater bryozoans. J. Morphology 179:197-202.

Mukai, Hideo, Takayoshi Niwa, Masashi Tsuchiya, and Takamasa Nemoto. 1983. The growth of colonies and the production of statoblasts in a freshwater bryozoan, *Plumatella casmiana*. Sci. Rep. Fac. Educ. Gunma Univ. 32:55-75.

Mundy, S. P. 1980. Stereoscan studies of phylactolaemate bryozoan statoblasts including a key to the statoblasts of the British and European Phylactolaemata. J. Zool. Lond. 192:511-530.

Mundy, S. P. and John P. Thorpe. 1979. Biochemical genetics and taxonomy in *Plumatella fungosa* and *P. repens* (Bryozoa: Phylactolaemata). Freshwater Biol. 9:157-164.

Oda, Shuzitu. 1955. Variability of the statoblast in *Lophopodella carteri*. Sci. Rep. Tokyo Kyoiku Daig. 8(B) (115):1-22.

Oda, Shuzitu. 1959. Germination of the statoblasts in freshwater Bryozoa. Sci. Rep. Tokyo Kyoiku Daig. 9(B) (135):90-131.

Oda, Shuzitu. 1960. Relation between sexual and asexual reproduction in freshwater Bryozoa. Bull. Mar. Biol. Sta. Asamushi 10(2):111-116.

Oda, Shuzitu. 1974. *Pectinatella magnifica* occurring in Lake Shoji, Japan. Proc. Jap. Soc. Syst. Zool. 10:31-39.

Oda, Shuzitu. 1979. Germination of the statoblasts of *Pectinatella magnifica*, a freshwater bryozoan. Pages 93-112, *in* Advances in Bryozoology. Gilbert P. Larwood and Marie B. Abbot, eds. Academic Press. London & New York. 639 p.

Oka, Asajiro. 1907. Zur Kenntnis der Süsswasserbryozoen von Japan. Annot. Zool. Jap., 6(2):117-123.

Oka, Asajiro. 1908. Ueber eine neue Gattung von Süsswasserbryozoen (*Stephanella* n.g.). Annot. Zool. Jap. 6:277-285.

Oka, Hidemiti and Shuzitu Oda. 1948. Observations on freshwater Bryozoa with special reference to their reproduction. Collecting and Breeding 10:39-48. (In Japanese).

Potter, Rosamond. 1979. Bryozoan karyotypes and genome sizes. Systematics Assoc. Spec. Vol. 13:11-32.

Potts, Edward. 1884. On *Paludicella erecta*. Proc. Acad. Nat. Sci. Philadelphia 36:213-214.

Rao, Kotapalli S. 1973. Studies on freshwater Bryozoa. III. The Bryozoa of the Narmada River System. Pages 529-537, *in* Living and Fossil Bryozoa: Recent Advances in Research. Gilbert P. Larwood, ed. Academic Press, London. 634 p.

Rao, Kotapalli S. and John H. Bushnell. 1979. New structures in binding designs of freshwater Ectoprocta dormant bodies (statoblasts). Acta. Zool. 60:123-127.

Rogick, Mary D. 1934. Studies on freshwater Bryozoa. I. The occurrence of *Lophopodella carteri* Hyatt 1866 in North America. Trans. Amer. Microsc. Soc. 53(4):416-424.

Rogick, Mary D. 1935a. Studies on freshwater Bryozoa. II. The bryozoa of Lake Erie. Trans. Amer. Microsc. Soc. 54(3):245-263.

Rogick, Mary D. 1935b. Studies on freshwater Bryozoa. III. The development of *Lophopodella carteri* var. *typica.* Ohio J. Sci. 35:457-464.

Rogick, Mary D. 1936. Studies on freshwater Bryozoa. IV. On the variation of statoblasts of *Lophopodella carteri.* Trans. Amer. Microsc. Soc. 55:327-333.

Rogick, Mary D. 1937a. Studies on freshwater Bryozoa. V. Some additions to Canadian fauna. Ohio J. Sci. 37(2):99-104.

Rogick, Mary D. 1937b. Studies on freshwater Bryozoa. VI. The finer anatomy of *Lophopodella carteri* var. *typica.* Trans Amer. Microsc. Soc. 56(4):367-396.

Rogick, Mary D. 1939. Studies on freshwater Bryozoa. VIII. Larvae of *Hyalinella punctata* (Hancock) 1850. Trans. Amer. Microsc. Soc. 58:199-209.

Rogick, Mary D. 1940. Studies on freshwater Bryozoa. XI. The viability of dried statoblasts of several species. Growth 4(3):315-322.

Rogick, Mary D. 1941a. The resistance of freshwater Bryozoa to dessication. Biodynamica 3(77):369-378.

Rogick, Mary D. 1941b. Studies on freshwater Bryozoa. X. The occurrence of *Plumatella casmiana* in North America. Trans. Amer. Microsc. Soc. 60(2):211-220.

Rogick, Mary D. 1943. Studies on freshwater Bryozoa. XIII. Additional *Plumatella casmiana* data. Trans. Amer. Microsc. Soc. 62(3):265-270.

Rogick, Mary D. 1945. Studies on freshwater Bryozoa. XVI. *Fredericella australiensis* var. *browni,* n. var. Biol. Bull. Woods Hole 89(3):215-228.

Rogick, Mary D. 1957. Studies on freshwater Bryozoa. XVIII. *Lophopodella carteri* in Kentucky. Trans. Kentucky Acad. Sci. 18(4):85-87.

Rousselet, C. F. 1904. On a new fresh-water Polyzoon from Rhodesia, *Lophopodella thomasi gen. et sp. nov.* Journ. Queckett Micr. Club 2:45-56.

Schröder, O. 1913. Uber einen einzelligen Parasiten des Darmepithels von *Plumatella fungosa* Pallas. Zool Anz. 43:220-223.

Shrivastava, Pradeep and Kotapalli S. Rao. 1985. Ecology of *Plumatella emarginata* (Ectoprocta: Phylactolaemata) in the surface waters of Madhya Pradesh with a note on its occurrence in the protected waterworks of Bhopal (India). Environ. Poll. (Ser. A) 39:123-130.

Sinclair, Ralph and Billy Isom. 1963. The occurrence of certain bryozoans in Tennessee waters. Tennessee Stream Poll. Bd. Publ. No. 10. 8 p.

Smith, Douglas G. 1985. *Lophopodella carteri* (Hyatt), *Pottsiella erecta* (Potts) and other freshwater Ectoprocta in the Connecticut River (New England, U.S.A.). Ohio J. Sci. 85(1):67-70.

Tenney, Wilton R. and William S. Woolcott. 1962. First report of the bryozoan *Lophopodella carteri* (Hyatt) in Virginia. Amer. Midl. Nat. 68:247-248.

Tenney, Wilton R. and William S. Woolcott. 1964. A comparison of the responses of some species of fishes to the toxic effect of the bryozoan, *Lophopodella carteri* (Hyatt). Virginia J. Sci. (n. ser.) 15:16-20.

Thorpe, John P. and S. P. Mundy. 1980. Biochemical genetics and taxonomy in *Plumatella emarginata* and *P. repens* (Bryozoa: Phylactolaemata). Freshwater Biol. 10:361-366.

Toriumi, Makoto. 1941a. Studies on freshwater Bryozoa of Japan. I. Sci. Rep. Tôhoku Imp. Univ. (4) 16(2):193-215.

Toriumi, Makoto. 1941b. Studies on freshwater Bryozoa of Japan. II. Freshwater Bryozoa of Tyosen (Korea). Sci. Rep. Tôhoku Imp. Univ. (4) 16(4):413-425.

Toriumi, Makoto. 1942a. Studies on freshwater Bryozoa of Japan. III. Freshwater Bryozoa of Hokkaido. Sci. Rep. Tôhoku Imp. Univ. (4) 17(2):197-205.

Toriumi, Makoto. 1942b. Studies on freshwater Bryozoa of Japan. IV. Freshwater Bryozoa of Taiwan (Formosa). Sci. Rep. Tôhoku Imp. Univ. (4) 17(2):207-214.

Toriumi, Makoto. 1951. Taxonomical study on freshwater Bryozoa. I. *Fredericella sultana* (Blumenbach). Sci. Rep. Tôhoku Univ. (4) 19(2):167-177.

Toriumi, Makoto. 1952a. Taxonomical study on freshwater Bryozoa. II. *Paludicella articulata* (Ehrenberg). Sci. Rep. Tôhoku Univ. 19(3):255-259.

Toriumi, Makoto. 1952b. Taxonomical study on freshwater Bryozoa, VI. *Plumatella emarginata* Allman. Sci. Rep. Tôhoku Univ. 19(4):320-334.

Toriumi, Makoto. 1955a. Taxonomical study on freshwater Bryozoa. X. *Plumatella casmiana* Oka. Sci. Rep. Tôhoku Univ. 21(4):67-77.

Toriumi, Makoto. 1955b. Taxonomical study on freshwater Bryozoa. XIII. *Hyalinella punctata* (Hancock). Sci. Rep. Tôhoku Univ. (4) 21(3-4):241-247.

Toriumi, Makoto. 1956. Taxonomical study on freshwater Bryozoa. XVI. *Lophopodella carteri* (Hyatt). Sci. Rep. Tôhoku Univ. (4) 22(1):35-44.

Toriumi, Makoto. 1962. Analysis of intraspecific variation in *Lophopodella carteri* (Hyatt) from the taxonomical view-point. I. Consideration on the variation of spine number of the spinoblast. Bull. Mar. Biol. Sta. Asamushi, Tôhoku Univ. 11(2):59-70.

Toriumi, Makoto. 1963a. Analysis of intraspecific variation in *Lophopodella carteri* (Hyatt) from the taxonomical view-point. II. Notes on the intraspecific groups with special reference to the spine number of the spinoblasts. Bull. Mar. Biol. Sta. Asamushi, Tôhoku Univ. 9(3):135-144.

Toriumi, Makoto. 1963b. Analysis of intraspecific variation in *Lophopodella carteri* (Hyatt) from the taxonomical view-point. III. Intraspecific groups distinguished by the dimensional character of the spinoblasts. Bull. Mar. Biol. Sta. Asamaushi, Tôhoku Univ. 9(3):145-152.

Toriumi, Makoto. 1963c. Analysis of intraspecific variation in *Lophopodella carteri* (Hyatt) from the taxonomical view-point. IV. Relation between the intraspecific groups and special form of the spinoblasts with numerous spines. Bull. Mar. Biol. Sta. Asamushi, Tôhoku Univ. 9(3):153-160.

Toriumi, Makoto. 1963d. Analysis of intraspecific variation in *Lophopodella carteri* (Hyatt) from the taxonomical view-point. V. Intraspecific groups distinguished by shape of the spinoblasts. Bull. Mar. Biol. Sta. Asamushi, Tôhoku Univ. 9(3):161-166.

Toriumi, Makoto. 1964a. Analysis of intraspecific variation in *Lophopodella carteri* (Hyatt) from the taxonomical view-point. VI. Intraspecific groups discriminated by the tentacle number. Bull. Mar. Biol. Sta. Asamushi, Tôhoku Univ. 12(1):13-20.

Toriumi, Makoto. 1964b. Analysis of intraspecific variation in *Lophopodella carteri* (Hyatt) from the taxonomical view-point. VII. Intraspecific groups distinguished by a relation between the tentacle and the spine numbers. Bull. Mar. Biol. Sta. Asamushi, Tôhoku Univ. 12(1):21-26.

Toriumi, Makoto. 1964c. Analysis of intraspecific variation in *Lophopodella carteri* (Hyatt) from the taxonomical view-point. VIII. Some observations on the distribution of the groups. Bull. Mar. Biol. Sta. Asamushi, Tôhoku Univ. 12(2-3):147-160.

Toriumi, Makoto. 1967a. Analysis of intraspecific variation in *Lophopodella carteri* (Hyatt) from the taxonomical view-point. IX. Additional observations on the variation of tentacle number. Bull. Mar. Biol. Sta. Asamushi. Tôhoku Univ. 13(1):13-20.

Toriumi, Makoto. 1967b. Analysis of intraspecific variation in *Lophopodella carteri* (Hyatt) from the taxonomical view-point. X. Further observations on the relation between the number of tentacles on the polypides and the number of spines on the spinoblast. Bull. Mar. Biol. Sta. Asamushi, Tôhoku Univ. 13(1):21-27.

Toriumi, Makoto. 1967c. Analysis of intraspecific variation in *Lophopodella carteri* (Hyatt) from the taxonomical view-point. XI. Prelimary observations on the after effect of temperature on the spine number of the spinoblast. Sci. Rep. Tôhoku Univ. (4) 33(3-4):487-498.

Toriumi, Makoto. 1971a. Additional observations on *Plumatella repens* (L.) (a fresh-water bryozoan). IV. Re-examination of the field materials of *P. repens* and *P. fungosa*. Bull. Mar. Biol. Sta. Asamushi, Tohôku Univ. 14(2):117-126.

Toriumi, Makoto. 1971b. Additional observations on *Plumatella repens* (L.) (a fresh-water bryozoan). V. Re-consideration on the relationship between *P. repens* and *P. fungosa* by the rearing. Bull. Mar. Biol. Sta. Asamushi Tôhoku Univ. 14(2):127-140.

Toriumi, Makoto. 1972. Additional observations on *Plumatella repens* (L.) (A fresh-water byrozoan). VII. Re-examination on the materials labeled *Plumatella punctata*. Bull. Mar. Biol. Sta. Asamushi, Tôhoku Univ. 14(3):155-167.

Toriumi, Makoto. 1974. Analysis of intraspecific variation in *Lophopodella carteri* (Hyatt) from the taxonomical view-point. XII. General consideration. Bull. Mar. Biol. Sta. Asamushi Tôhoku Univ. 15(1):1-12.

Trauberg, O. 1940. Beitrag zur Kenntnis einiger in Lettland vorkommender Süsswasser-Bryozoan-Arten. Folia Zool. Hydrobiol. 10(2):479-484.

Viganò, Antonio. 1968. Note su *Plumatella casmiana* Oka (Bryozoa). Riv. Idrobiologia 7(3):421-468.

Wiebach, Fritz. 1973. Preliminary notes on a revision of the genus *Hyalinella*. Pages 539-547, *in* Living and Fossil Bryozoa: Recent Advances in Research. Gilbert P. Larwood, ed. Academic Press, London & New York. 634 p.

Wiebach, Fritz. 1974. Specific structures of sessoblasts (Bryozoa: Phylactolaemata). Docum. Lab. Geol. Fac. Sci. Lyon H. S. 3(1):149-154.

Wood, Timothy S. 1973. Colony development in species of *Plumatella* and *Fredericella* (Ectoprocta: Phylactolaemata). Pages 395-432, *in* Development and Function of Animal Colonies Through Time. Richard S. Boardman, Alan H. Cheetham, and William A. Oliver, Jr., eds. Dowden, Hutchinson and Ross, Stroudsburg, Pennsylvania. 601 p.

Wood, Timothy S. 1979. Significance of morphological features in bryozoan statoblasts. Pages 59-73, *in* Advances in Bryozoology. Gilbert P. Larwood and Marie B. Abbott, eds. Academic Press, London & New York. 639 p.

Wood, Timothy S. 1988. *Plumatella reticulata* sp. nov. in Ohio (Bryozoa: Phylactolaemata). Ohio J. Sci. 88(3)101-104.

Zimmerman, Michael W. 1979. Larval release and settlement behavior in three species of plumatellid Bryozoa (Ectoprocta). M. S. Thesis, Wright State Univ., Ohio. 52 p.

GLOSSARY

ancestrula. A single zooid which develops from a statoblast or which is established (usually with others) directly from a larva; it may have fewer tentacles than the species average, but otherwise it is anatomically identical to subsequently budded zooids (Figure 2c).

annulus. That portion of the periblast which surrounds the fenestra; in floatoblasts the annulus is usually composed of gas-filled chambers, in sessoblasts it is a thin, transparent flange called a lamella.

budding. Any form of asexual reproduction in which tissues from the adult redifferentiate to form a new individual; in the case of bryozoans, budding includes the formation of statoblasts as well as the direct outgrowth of daughter zooids from parental tissues.

Bryozoa. The phylum name once used to designate the various marine and freshwater "moss animals;" synonymous with Polyzoa and Ectoprocta, the latter now preferred to distinguish between the ectoprocts and entoprocts (see Ectoprocta).

bryozoan. General common name for an animal classified in the Phylum Ectoprocta, sometimes also used loosely to include unrelated animals in the Phylum Entoprocta; (plural : bryozoans, not bryozoa).

capsule. Inner core of a statoblast containing germinative material for a new zooid, generally surrounded by two halves of a protective periblast (Figure 4).

capsuled floatoblast. Any floatoblast bearing an internal capsule, which for some time contains dormant germinative tissues; the term is normally used to distinguish this typical floatoblast from the unusual thin-walled leptoblast of *Plumatella casmiana* (Figure 15c).

coelom. Fluid-filled body cavity surrounded by tissues derived from mesoderm; in bryozoans, the coelom includes the main body cavity in which all polypides are suspended (Figure 1a); coelomic spaces also penetrate the lophophore and each hollow tentacle.

colony wall. See cystid.

cystid. Composite tissues and secreted materials which separate the coelomic (body) cavity from the surrounding water, also called a zooecium; includes peritoneum, muscle fibers, dermal tissues, and a nonliving ectocyst (Figure 1a).

dorsal valve. One of a pair of floatoblast valves which bears the largest area of gas-filled chambers; because of its greater buoyancy, this valve is most likely to face upwards when the statoblast is floating (Figure 3a).

ectocyst. Nonliving material secreted as an external layer of the colony wall, ranging in texture from sclerotized or membranous to gelatinous and slimy.

Ectoprocta. Phylum of colonial marine and freshwater animals characterized by a ciliated lophophore encircling a mouth, and an anus located outside the whorl of tentacles; commonly known as bryozoans.

emargination. A condition found in certain encrusted, branching colonies in which the transparent tip of a zooid is clearly demarcated from the surrounding opaque ectocyst (Figure 21a; 23a).

epistome. An active lobe of tissue which overhangs and partially blocks the mouth opening in Phylactolaemate bryozoans; assumed to function in food particle selection (Figure 1a).

fenestra. The clear, central area of a periblast valve which reveals the underlying statoblast capsule (Figure 4).

floatoblast. A statoblast having an annulus of gas-filled chambers which provide buoyancy; in some cases, the annulus may become buoyant only after it is dried, eg. *Hyalinella* and *Lophopodella* (Figure 3a-c).

fragmentation. Asexual reproduction by the spontaneous breaking of the individual (colony) into viable parts.

funiculus. A tubular strand of tissue joining the blind end of the gut to the colony wall; instrumental in the production of statoblasts and spermatozoa (Figure 1a).

furrow. A thin, transparent line occurring along the crest of the keel in branching, tubular colonies (Figure 15a).

Gymnolaemata. The largest of three classes of living Ectoprocta, distinguished from Phylactolaemata by the absence of statoblasts, lack of an epistome, small size, circular lophophore, and a mainly marine habitat.

hibernaculum. An irregularly-shaped resting bud occurring in certain freshwater gymnolaemate bryozoans, appearing as an enlarged portion of the zooecium.

incrustation. A layer of particles adhering to the ectocyst of certain branching, tubular bryozoans rendering the colony wall nearly opaque.

keel. A raised, longitudinal ridge extending along the attached, tubular portions of certain colonies, varying from prominent to barely perceptible (Figures 21a; 23a).

lamella. A thin, flange-like peripheral extension of the dorsal periblast in sessoblasts, homologous to the annulus of floatoblasts.

lamella. See annulus.

larva. Normally, an immature motile phase structurally unlike the adult. In the case of phylactolaemate bryozoans, the larva contains at least two fully-formed polypides enclosed in a heavily ciliated mantle (Figure 1b).

leptoblast. An unusual, thin-walled statoblast which bypasses diapause and lacks the resistance to dessication normally found in other statoblasts; formed only in *Plumatella casmiana* (Figure 15b).

lophophore. An organized structure of ciliated tentacles used for capturing suspended particles in the water and probably also a major site for the exchange of respiratory gases (Figure 1a).

orifice. The terminal opening in a zooid through which the polypide is extended.

periblast. The shell-like structure forming a protective covering around the capsule in most statoblasts; formed from paired valves, each with a clear central fenestra and a peripheral annulus or lamella (Figure 4).

Phylactolaemata. The smaller class of the Phylum Ectoprocta, characterized by an epistome, statoblast formation, usually a horseshoe-shaped lophophore, and an exclusively freshwater habitat.

pipotoblast. A simple type of statoblast, apparently lacking a periblast, sometimes with ventral, keel-like projections which grip that part of the ectocyst attached to the substrate; formed only in the family Fredericellidae (Figure 3f-g).

polypide. The retractable portion of a zooid comprising the central ganglion and all structures related to feeding and digestion (Figure 1a).

septum. In bryozoans, an incomplete internal partition sometimes occurring in tubular colonies, especially in *Plumatella emarginata* and *P. reticulata* (Figure 21b); its function is unknown.

sessoblast. A relatively large statoblast cemented firmly through the colony
wall to the substrate; formed only in the family Plumatellidae (Figure
3d-e).

statoblast. A specialized bud produced by all phylactolaemate bryozoans,
consisting of germinal tissue and yolky material encapsulated in a chitinous shell, plus any additional protective outer structures (see p. 5).

suture. The area of a statoblast where the two valves are jointed.

valve. One of a pair of sclerotized structures forming the outer covering of
a statoblast, each composed of a capsule and a periblast (Figure 4).

ventral valve. That one of a pair of floatoblast valves which bears the
smallest area of gas-filled chambers; because of its relatively weak buoyancy, this valve tends to face downward when the statoblast is floating
(Figure 3b).

vestibular pore. A temporary opening near the orifice in a zooid through
which the leptoblast is released in *Plumatella casmiana.*

zooecium. See cystid.

zooid. One of the physically connected, asexually replicated, morphologic units
that comprise a colony; in bryozoans the zooid includes the polypide and
associated musculature, plus all parts of the adjacent cystid (Figure 1a).